Digital Afterlife and the Spiritual Realm

Artificial Intelligence and Robotics Series
Series Editor: Roman Yampolskiy

Artificial Intelligence Safety and Security
Roman V. Yampolskiy

Artificial Intelligence for Autonomous Networks
Mazin Gilbert

Virtual Humans
David Burden, Maggi Savin-Baden

Deep Neural Networks
WASD Neuronet Models, Algorithms, and Applications
Yunong Zhang, Dechao Chen, Chengxu Ye

Introduction to Self-Driving Vehicle Technology
Hanky Sjafrie

Digital Afterlife
Death Matters in a Digital Age
Maggi Savin-Baden, Victoria Mason-Robbie

Multi-UAV Planning and Task Allocation
Yasmina Bestaoui Sebbane

Cunning Machines
Your Pocket Guide to the World of Artificial Intelligence
Jędrzej Osiński

Autonomous Driving and Advanced Driver-Assistance Systems
Applications, Development, Legal Issues, and Testing
Edited by Lentin Joseph, Amit Kumar Mondal

Digital Afterlife and the Spiritual Realm
Maggi Savin-Baden

For more information about this series please visit: https://www.routledge.com/Chapman--HallCRC-Artificial-Intelligence-and-Robotics-Series/book-series/ARTILRO

Digital Afterlife and the Spiritual Realm

Maggi Savin-Baden

CRC Press
Taylor & Francis Group
Boca Raton London New York

CRC Press is an imprint of the
Taylor & Francis Group, an **informa** business

A CHAPMAN & HALL BOOK

First edition published 2022
by CRC Press
6000 Broken Sound Parkway NW, Suite 300, Boca Raton, FL 33487-2742

and by CRC Press
2 Park Square, Milton Park, Abingdon, Oxon, OX14 4RN

© 2022 Maggi Savin-Baden

CRC Press is an imprint of Taylor & Francis Group, LLC

ISBN: 978-0-367-56538-1 (hbk)
ISBN: 978-0-367-56462-9 (pbk)
ISBN: 978-1-003-09825-6 (ebk)

DOI: 10.1201/9781003098256

Typeset in Minion
by Deanta Global Publishing Services, Chennai, India

In the universe, there are things that are known, and things that are unknown, and in between them, there are doors.

(WILLIAM BLAKE, 1794)

For, John, see you there, later

Contents

List of Figures
and Tables

TABLES

FIGURES

Acknowledgements

This is a book that has shifted, grown and moved over the time of writing thanks to the many discussions I have had with people during this time. I am grateful in particular to all the participants who took part in the study and whose reflections were used not only in Chapter 9 but also throughout the book.

Thanks are due particularly to Rev Dr Jeremy Brooks and Rev Dr John Reader, who were kind enough to read and comment on the whole manuscript. Thanks are also due to Amolak Dhillon, Professor Jaswinder Dhillon, Professor Liz Falconer, Dr Becca Khanna, Raj Khanna, Umrah Mahadik, Rabbi Jason Miller and Professor Richard Woolley, who read and commented on Chapter 2, and to Randi Cohen at Taylor and Francis, who has been so supportive in seeing it through.

I am also deeply thankful to my husband John, who has supported the editing process and in a home of structured chaos is always there to steady the buffs. Any mistakes and errors are mine.

Permissions

Crazy Coffins provided permission for the use of the screenshot (Figure 2.1) in Chapter 2.

In The Light Urns, Inc. provided permission for the use of the screenshot (Figure 2.2) in Chapter 2.

And Vinyly provided permission for the use of the screenshot (Figure 2.3) in Chapter 2.

Celestis, Inc. provided permission for the use of the screenshot (Figure 2.4) in Chapter 2.

James Norris provided permission for the use of the screenshot (Figure 10.1) of the MyWishes site in Chapter 10.

Philip Stone kindly gave permission for the use of his DT Spectrum figure (Figure 11.1) in Chapter 11.

Introduction

F EW RELIGIOUS LEADERS HAVE examined the potential for the positive impact of digital media and digital immortality creation in religious contexts. Marking the end of life today increasingly includes memorials, whether sites of roadside crashes, family and friends 'saying a few words' or listening to the deceased's favourite music during religious services. It is evident that there have been recent moves away from traditional funeral services focusing on the transition of the deceased into the future world beyond, towards a rise of memorial content within funerals and commemorative events, which herald shifts in afterlife beliefs by replacing them, to all intents and purposes, by attitudes to this life. Irrespective of religious beliefs, most people present at memorials can reflect on the deceased person's life and find something worth recalling. It is a shared act that induces a sense of singleness of purpose without introducing divisive beliefs. In a post-modern context of mixed religious beliefs and secular outlooks, this affords a safe ritual space. Now, digital afterlife represents the next quantum leap in creating potentially everlasting memorials. Further, it is not clear whether the possibility of digital immortality and the use of digital media alter thoughts about the mind–body connection and whether interaction with a digital immortal alters one's spiritual journey through grief. This sense of investment in flesh results in a perception that what happens in the digital context is disembodied and therefore less real, less incarnational, less faith related. One of the central difficulties is that many people find digital afterlife creation disconcertingly disembodied and it is not clear whether it promotes particular views about bodily forms in the afterlife. Furthermore, digital afterlife creation may prevent us from being free to die (Davies, 2008: 106), as well as introducing questions about how 'the dead' might be classified.

The book aims to

a) Examine how digital media and the creation of digital immortals may affect religious understandings of death and the afterlife within a religious context

b) Understand the impact of digital media on those living and those working with the bereaved

c) Explore the impact of digital memorialization post-death

d) Understand the ways in which digital media may be changing conceptions and theologies of death

The book begins with Chapter 1, 'Digital Afterlife and the Spiritual Realm', by exploring a range of views about digital afterlife and the idea of the spiritual realm in relation to this. It then examines depictions of the spiritual realm in terms of places, spaces and relationships. Chapter 2, 'Religion and Afterlife', provides a more in-depth exploration of the issues and presents perspectives on death and the afterlife in relation to six different belief systems, namely Christianity, Humanism, Sikhism, Buddhism, Hinduism and Islam. It draws on interviews and email conversations from those from particular faiths in order to compare diverse and overlapping perspectives on the afterlife, and current trends in death merchandise and grave goods. The practical concerns are presented in Chapter 3, 'Impact of the Digital on the Bereaved and Bereavement', by exploring models of death and grief, and analyzing the unintended consequences that the array of digital practices may have on the bereaved and ways in which the bereaved might be supported better in digital spaces.

Chapter 4, 'Digital Theologies and Death', examines digital theologies and how they have been defined, adopted and implemented and then seeks to bring some clarity to these ideas, as well as exploring religious perspectives on media use in relation to theology. The final section of the chapter explores digital theology in relation to digital afterlife, offering perspectives from a study that examined perceptions of digital afterlife. Chapter 5, 'Death Spaces', focuses on physical spaces such as cemeteries, memorial sites, crematoria and burial grounds, and then digital spaces including online cemeteries and online funerals. The final sections of the chapter explore digital death spaces and new death spaces, suggesting that death spaces are ones that need to be interrogated continually.

There have been many discussions about the idea of a good and a bad death and in Chapter 6, a good digital death is explored. It is argued that planning, preservation, mediation, transference and taboo management are central concepts in the administration of a good digital death and ways of achieving this are suggested. The second section of the chapter presents areas that can affect not only the possibility for a good death but also good mourning in terms of taboos, etiquette and unintended consequences. The complexity of unintended consequences is explored further in Chapter 7, 'Death and the Liminal', which couples the research on liminal spaces with perspectives of death and experiences of bereavement. It is argued that spiritual spaces are often liminal and feel other-worldly and they prompt contemplation about the possibility of worlds beyond and in particular the afterlife. Chapter 8, 'Symbols and Memorialization', builds on the liminal by analyzing the use of the symbols associated with death and the way in which they are transposed into digital spaces. It then examines the relationship between semiotics and spirituality as well as the purpose and practice of memorialization.

One of the difficulties in the research and literature to date is that there are few studies that have explored digital afterlife and the spiritual realm. Chapter 9, 'Perspectives on Digital Afterlife', presents the findings from a study that examined how digital media and digital afterlife creation affected understandings of death and the afterlife within religious contexts. It suggests that we are currently experiencing a shift from the digital to the postdigital and this is resulting in postdigital theologies.

The final three chapters of the book explore new and current research and practices, beginning with Chapter 10, 'Digital Legacy'. This chapter explores legal and ethical issues, presents the recent research and literature on digital legacy, and then explores issues of ownership and privacy. It also examines digital estate planning and the death tech industry. Chapter 11, 'Ambivalence and Spectacle' develops this by examining some of the more recent macabre practices. It explores the idea of spectacular death and suggesting that the use of artificial intelligence has prompted the development of spectacular death as well as an ambivalence towards it. The final part of the chapter explores the growth and change and impact of dark tourism, including the rise of immersive dark tourism and the changing landscape of thanatopathia and thanatopolitics. The last chapter, 'The Final Cut' explores the relationship between physical and digital death using the metaphor of the final cut. It then examines the impact of absent presence on the living and then the notions of the digital death pragmeme. Finally, it reflects on the ongoing confusion and conundrums related to digital afterlife and the spiritual realm.

This book explores the ways in which digital media and digital afterlife affect social and religious understandings of death and the afterlife. It examines the impact of digital media on concepts of death in contemporary society and discusses digital memorialization post-death and its impact on the bereaved. Digital afterlife has moved beyond digital memorialization towards a desire to preserve oneself after death. Preserving oneself or being preserved by someone else may affect both the dying person's peace of mind and the well-being of the bereaved. Digital media are currently being used to expand the possibilities of commemorating the dead and managing the grief of those left behind, complementing and sometimes replacing the well-established formal structures and rituals of religion and faith. For many people, digital afterlife and the spiritual realm largely remains an area that is both inchoate and confusing. This book aims to begin to unravel some of this bafflement.

CHAPTER 1

Digital Afterlife and the Spiritual Realm

INTRODUCTION

Afterlife, digital afterlife and the spiritual realm tend to be couched in mystery. Philosophical and theological discourse cannot explain the resurrection or reincarnation of the body, because the human body itself is not reducible to a simple description or ready comprehension. New practices suggest that the possibilities of 'living on' through technological innovations have the potential to change the religious landscape radically. Recent developments suggest that there will be socio-political and psychological effects that will have an impact on understandings of embodiment and death, and create new forms of post-mortem veneration. This chapter explores a range of views about digital afterlife and the idea of the spiritual realm in relation to this. It begins by examining the concept of symbolic immortality and then presents the diverse interpretations of digital afterlife. The second section of the chapter explores depictions of the spiritual realm in terms of places, spaces and relationships.

OVERVIEW

Whilst there has been some research into digital afterlife perceptions about the ways in which digital media affect how we see living and dying remain complex areas for research and discussion. It is also not clear whether changed perceptions of living and dying in the light of digital media practices are affecting religious views about illness and notions of

DOI: 10.1201/9781003098256-1

1

afterlife, heaven, hell and salvation. In the 21st century, death is integrated into life for many people through social media so that the dead reside in our machines and phones. Recent developments seem to suggest shifts in understandings about embodiment, death and afterlife (Walter, 2017). For example, digital media are currently being used to expand the possibilities of commemorating the dead and managing the grief of those left behind, complementing and sometimes replacing the well-established formal structures of faiths and belief systems.

The concept of digital afterlife is defined here as the continuation of an active or passive digital presence after death (Savin-Baden et al, 2017). Other terms have been used to describe digital afterlife including digital immortality. 'Afterlife' assumes a digital presence that may or may not continue to exist, whereas 'immortality' implies an everlasting presence. The area of digital afterlife research is a growing field, and work initially began by exploring ways in which the dead were seen to be kept alive by the living. For example, Howarth (2000) presented ways in which afterlife was being managed over 20 years ago such as anniversaries, the creation of self-help groups. Other forms of communion and communication with the dead have been in use for a long time, such as talking to a loved one at their grave as well as spiritualism and clairvoyance. Examples also include what is now referred to as a durable biography (Walter, 1996) that allows survivors to continue to integrate the deceased person into their lives and to find a stable and secure place for them. More recent phenomenon includes cenotaphization (Kellaher and Worpole, 2010), whereby the remains are dislocated from places of memorializing (discussed in more depth in Chapter 8) and the creation of dynamic biographies. Dynamic biographies are when parents create biographies for deceased children, often through life stages and sometimes building a portrait of their achievements (Hockey, 1996). Amidst this varied landscape, the digital has begun to overlay many of the current physical practices, some of which sit side by side, whilst others replace grave spaces as ways of reconstituting the dead. What we see in cyberspace is the collision of worlds of the dead and living which not only overlap but tend to collide with one another and appear to offer misplaced hope, as Mosco argues:

> The thorny questions arising from all the limitations that make us human were once addressed by myths that featured gods, goddesses, and the variety of beings and rituals that for many provide satisfactory answers. Today, it is the spiritual machines and their

world of cyberspace that hold out the hope of overcoming life's limitations.

(Mosco, 2004: 78)

The debates in the field of digital afterlife are complex and wide ranging. Whilst perspectives in the 2000s tend to focus on robotics, memorialization and the creation and maintenance of digital beings, the 1970s focused more on the nature of immortality.

FORMS OF IMMORTALITY

Symbolic immortality (Lifton, 1973) is the idea that individuals seek for a sense of life continuity, or immortality, through symbolic means. This term was used by Lifton (1973) to describe ways of avoiding death through four different ways, namely biological, social, natural and theological. However, symbolic immortality could also include the concepts of assisted immortality (Kastenbaum, 2004) and one- and two-way immortality (Bell and Gray, 2000).

Biological immortality is the belief that through transmitting our genes via our descendants we continue. The idea is that family heritage continues both genetically as well as by passing on values, philosophies and memories from generation to generation. Thus, there is a sense that someone lives on physically – and possibly spiritually, through one's children and grandchildren.

Social immortality is the idea that we live on by creating artefacts or creations that survive us, such as books, arts or even the influences we may have had on friends or students. Thus, we live on beyond death through artefacts we have created or acts we have undertaken – such as benevolence, so that we will be remembered for generations and possibly centuries.

Natural immortality is the recognition that as one's body returns to the ground it becomes part of the earth's life cycle. Thus, our bodies, returned to the earth become part of the life and death cycle of nature.

Theological immortality is the immortalization of the soul after death and, as will be discussed in Chapter 2, is central to a number of different religions. The afterlife, with an immortal soul, is an ancient mythological theme involving death, rebirth and resurrection. Life after death, however, is not a traditional view in Jewish or humanist religious philosophies.

Assisted immortality was introduced in 2004 by Kastenbaum to capture the idea of technology-assisted survival. His proposition was how people could delineate what might be a meaningful form of survival if they made use of any available technological assistance.

One- and two-way immortality was established by Bell and Gray (2000). One-way immortality is where someone's ideas and digital profile have been preserved or memorialized. Two-way immortality is the idea that there is the potential for the creator to interact with the living world; this interaction could come in a wide variety of ways, from two-way text or even voice and video conversations by creating a robot that accessed previous posts and text messages.

There has been a shift away from the term 'immortality' towards the broader and more inclusive term, 'afterlife'. The notion of afterlife includes a wide variety of ideas and practices, as presented in Table 1.1.

DIGITAL AFTERLIFE AND THE SPIRITUAL REALM

The spiritual realm is defined here as a realm that is connected to the physical world. It exists alongside. Many people see the spiritual realm as something akin to heaven or somewhere we go post-death; others see it as a place of ghosts, spirits and deities. Most religions suggest that there is a body, a soul and a spirit, with the soul consisting of the intellect, will and emotions, whilst the spirit is more ethereal and is the component that interacts with the spiritual realm, communes with God and also includes things like conscience or intuition. Perhaps the most useful way of seeing the spiritual realm is that provided by Hick who argued that we need to be aware of the spiritual realm, 'the fifth dimension' (the other four being three dimensions of space and one dimension of time). He builds on Kant's distinction between the noumena and phenomena, the idea that the phenomenal world is the world we are aware of; this is the world we construct out of the sensations that are present in our consciousness. The noumenal world consists of things we seem compelled to believe in but which we can never know. This is because although the noumenal holds the contents of the intelligible world, Kant (1781/2007) claimed that we can only know phenomena and can never penetrate to the noumenon. Despite this, we are not entirely excluded from the noumenal because of the ability to act as a moral agent; this makes no sense unless a noumenal world exists as a place in which freedom, God and immortality abide. Hick suggests that:

TABLE 1.1 Features of Digital Afterlife (Adapted and Developed from Savin-Baden and Mason-Robbie, 2020)

Term	Definition	Example/s	Related research
Digital traces	Digital footprints left behind through digital media	Playlists Blog posts Website searches	Mayer-Schonberger (2009)
Digital legacy	Digital assets left behind after death	Things that are static once the user has died	Maciel and Pereira (2013)
Digital death	Either the death of a living being and the way it affects the digital world or the death of a digital object and the way it affects a living being	The impact of left-behind digital traces on family or the need for people to delete digital media because of its impact on everyday life	Pitsillides, Waller and Fairfax (2012)
Digital afterlife	The idea of a virtual space, where information, assets, legacies and remains reside as part of the cyber soul	The platform Dead Social enables people to schedule posts after they have passed away. Not really a virtual heaven or a place for souls to reside but it creates some of the illusion of this	Bassett (2018a)
Digital remains	Digital content and data which were accumulated and stored online during our lifetime that reflect our digital personality and memories	They are not uniform: these remains can be intangible assets, intellectual property, information about physical or tangible property, or personal data	Birnhack and Morse (2018)
Posthumous personhood	The idea of a model of a person that transcends the boundaries of the body	Someone being able to tweet or be reanimated through an avatar	Meese et al. (2015)
Second death	The deletion of digital remains	The grief experienced if/when someone's social media profile is deleted (by a social media platform) after death without consulting friends and family	Stokes (2015)
Second loss	The loss experienced due to the deletion of digital remains	People not upgrading their phones because they are afraid to lose voice messages from dead loved ones	Bassett (2018b)

(Continued)

TABLE 1.1 (CONTINUED) Features of Digital Afterlife (Adapted and Developed from Savin-Baden and Mason-Robbie, 2020)

Term	Definition	Example/s	Related research
Technologically mediated mourning	The use of social networking sites to mourn and memorialize those who have died physically	The setting up of blogs and social media for the purposes of grief management sites after events such as the Manchester Arena bombing in the UK in 2017 and numerous school and college shootings in the US	Kasket (2012)
Digital endurance	The creation of a lasting digital legacy and being posthumously present through digital reanimation	Using sites such as SocialEmbers or the KeepTheirMemoryAlive mobile phone app or an existing system which lives on after their death, such as a person's in-life profile that has been put into memorialized/remembering status in services such as Facebook	Savin-Baden and Burden (2019)
Digital resurrection	The use of dead people in media after death	Oliver Reed's role in *Gladiator* was finished using a digitally constructed face mapped onto a body double during editing since he died just before the final shoot of the film	Sherlock (2013)
Digital necromancy	The preservation and reanimation of digital remains	Deceased rapper Tupac Shakur performing live on stage with Snoop Dogg	Sherlock (2013)
Digital inheritors	Those who inherit digital memories and messages following the death of a significant other	People who expectedly or unexpectedly 'inherit' the social media profiles and messages following the death of close friends and relatives and who may or may not be able to delete them	Bassett (2018a)

(Continued)

TABLE 1.1 (CONTINUED) Features of Digital Afterlife (Adapted and Developed from Savin-Baden and Mason-Robbie, 2020)

Term	Definition	Example/s	Related research
Digital mourning labour	This is an activity undertaken by corporate brands who use social media to share (and gain from) emotions of grief and nostalgia about dead celebrities	After the death of David Bowie, the music and fashion industry shared their grief on social media using images such as the thunderbolt, the signature sign of Bowie	Kania-Lundholm (2019)
Augmented eternity	Bridging the gaps between the life by eternalizing digital identity	Using algorithms to answer questions from beyond the grave to create new forms of intergenerational intelligence	Tynan (2016)
Digital consciousness	The development of emotional and intellectual immortality through mind clones	Replication of mind-based memories, feelings and beliefs, exemplified in the LifeNaut project	Rothblatt (2014)

the critical realist principle-that there are realities external to us, but that we are never aware of them as they are in themselves, but always as they appear to us with our particular cognitive machinery and conceptual resources.

(HICK, 2004: 47)

What is central to this view of the spiritual realm is that religions should not just be seen as a set of beliefs but instead as spaces of unity with God/ Ultimate Reality, including engagement with issues that encompass social and political justice.

Religions across the world have been one of the main ways of helping people to cope with and make sense of death. What is interesting about the discipline of death studies, as well as the more recent research into digital afterlife is that they remain largely unconnected with the spiritual side of death and bereavement. However, the idea of hope is evident in some of the work in this area, such as that of Kubler Ross (1969) and Seale (1998).

Hope is central to many religions, but it is also evident in the language and practices of those with little or no belief system. Hope, it would seem, brings a sense of the spiritual into death spaces and appears particularly relevant to the idea of digital afterlife since it focuses on the human need to see beyond immediate grief. Hope is evident in a variety of ways in secular society, such as

- Experiencing the sense of the person after they have died
- A secular belief in some kind of life after death
- The hope brought by burying the dead in a woodland burial site providing the expectation of them becoming part of the environment and living on through the environment
- Placing cremated remains in places that connect with the life and identity of the deceased
- The hope that burying the dead in a beautiful place will be appreciated by the deceased

This kind of hope tends to centre on the physical positioning of the body post-death and for some people, cryogenics also offers a sense of hope. The preservation of the deceased through the digital also brings hope, through new layers of complexity rather than the expectations of hope inherent in the physical positioning of the body. There are assumptions that preserving oneself or being preserved by someone else may affect both the dying person's peace of mind and the well-being of the bereaved. Yet it is not clear whether this is linked to any kind of spiritual reasoning or hope, or whether it is seen as a form of comfort for the bereaved or indeed whether the deceased feel they need to leave behind something of significance. The motivation for creating any kind of digital afterlife is obscure. Further, it is not clear whether the possibility of a digital afterlife and the use of digital media alters thoughts about the mind–body connection, and whether interaction with a person's digital afterlife alters one's spiritual journey through grief. Distinguishing between those who preserve their own or another's digital afterlife, those that mediate the experience of others, and those that receive digital afterlife of a deceased person is important when considering the relationship with the digital realm. It is suggested here that there are three forms of afterlife creation. First, there are memory creators who use digital memories and artefacts pre and post the subject's death created by those left behind. Second, there are those who create a representative interactive avatar of themselves predeath which is able to conduct a limited conversation with others but has a very limited capability to learn, grow, act on and influence the wider world around it. The final type is persona creators, those people who choose to create a digitally immortal persona predeath that can learn and adapt over time and ultimately will be able to exert influence on the wider world around it. Although the first types are possible, the final forms of creation would be classed as virtual humans with sentience and as yet this is not possible. What is common to all of these forms of creation is that they tend not to include any acknowledgement of the spiritual realm.

PLACES, RELATIONSHIPS AND THE SPIRITUAL REALM

In many of the 21st-century textbooks on death, there is a notable absence of any reference to religion or spirituality. Yet from a spiritual point of view, digital memorial sites have become spaces of remembrance, consolation and condolence and to a degree these sites have replaced or recreated ritual practices associated with mourning. Invariably, discussions about spirituality, healing and peace occur in blogs today rather than books, particularly end-of-life bloggers. Further, what is meant by spirituality and the spiritual realm varies between those with strong religious affiliation and belief systems and those who just feel that the afterlife must bring some kind of peace after death.

The spiritual realm whilst existing as a realm, alongside the physical, is also seen here as something that is space, place and relationship. There has been considerable discussion about the nature of space and place. Space is defined here as something that is abstract, a dimension rather than the physical entity that would be described as place. It is a product of interaction; it continually shifts in meaning and is continually under construction and reconstruction. Thus, a place is seen as a location, something that is occupied, whilst space is a location that is seen as something more general. However, it is important to recognize that space and place are not always clearly distinguished from one another or their meaning is reversed (as in de Certeau, 1984), as Lammes argues:

> According to de Certeau these two conceptions of spatiality (maps and tours) are both incongruous dimensions of contemporary culture: we are confronted with a static representation of the world we live in, while at the same time sensing our space in a dynamic and more personal way. As place and space, maps and tours necessitate one another and come into being through two-way movement. Even more, a map always presupposes a tour, since one first needs to go somewhere to give an objective spatial account of it (de Certeau, 1984: 117–21).
>
> (Lammes, 2008: 87–88)

Thus, spiritual places have a sense of journeying to – with or without a map, as well as a sense of arriving and departing.

Spiritual Places

The idea of space brings with it a location; we see the world from a place and thus a place is always contextually related. For example, Retsikas (2007, 971–2) argues that a place is a tool of sociality because of:

> the changing character of place and the changing character of people's relationships with one another in, through, and, most importantly, with the places they find themselves as they move and stop, settle, and move again. In this sense, places are shifting and changing, always becoming through people's engagements

– material as well as discursive – in, through, and with them. As forms of social relationships are constantly deployed and realized spatially, places cannot be thought of as the mere geographical settings for social action.

Spiritual places are invariably seen as sacred places because they perform a religious function, not necessarily because they have peculiar physical or aesthetic qualities. Such spaces evoke grief and sadness through memories, particularly if a loved one dies in a specific place such as the roadside, yard or allotment. However, in the context of digital afterlife spiritual places are often spoken of as places with a particular meaning, as places in which the dead dwell or wait. For example, in the Catholic tradition of Christianity, there is a belief in purgatory, a place where souls go to be purified or to receive temporary punishment in order to make them ready before they are resurrected to heaven. Heaven is seen in the Christian tradition as where God resides and it is also seen as a place where renewed bodies and souls go after death. Heaven is described both as a place where God dwells and as a place that comes down to earth when Christ returns and there will be a new heaven. Gooder (2011) explains that a heaven is a place that exists alongside earth, not a place that is above us in the sky. The opposing option is Hell, where the wicked await resurrection before they find out their fate, or the place that is the eternal fate of the wicked which involves fire and possible destruction in eternal flames. However, grief is seen by some to be a sacred place. Rohr (2002) suggests that grieving is a sacred space, and it is a liminal space. He suggests that such liminal spaces are unique spiritual spaces where human beings hate to be but where the biblical God is always leading them. The reason he suggests that humans dislike such spaces is that they challenge us to live with mystery and paradox; they are spaces we cannot control yet at the same time they are the ones in which alternative consciousness can emerge, something that stretches beyond self-interest and self-will. Such sacred spaces also extend into spiritual relationships with people.

Spiritual Relationships

Spiritual relationships are defined as relationships with people who help to guide the spiritual management of bereavement, the relationships with spiritual beings whilst we are on this earth as well as the relationships with those who have died.

Relationship with people tends to focus on religious leaders and funeral directors as well as those who help to mediate the bereavement process. In the main, relatively few religious leaders are currently offering advice and guidance on any kind of digital afterlife creation. Some funeral directors do provide some afterlife services such as memorial sites and light online candle services. On the whole, it is private companies who are the ones experienced in the death tech industry. Such companies often have a broad view of what is possible and available and seek to ensure an understanding of the wishes and desires of the person predeath and the potential idiosyncratic response of the receivers to experiencing digital afterlife of the deceased person. Early examples included Eter9 which describes itself as 'is a social network that relies on artificial intelligence as a central element' and that 'Even in your absence, the virtual beings will publish, comment and interact with you intelligently'. Another site is Eternime whereby the individual train their immortal prior to death through daily interactions; but both Eter9 and Eternime in 2021 seem to be dormant. In many death tech sites, it is suggested that data are mined from Facebook, Fitbit, Twitter, e-mail, photos, video and location information with the individual's personality being developed by algorithms through pattern matching and data mining. A site that offers predeath planning services is MyWishes (previously called DeadSocial) which also offers digital legacy and digital end-of-life planning tutorials. A recent site is Aura, whose vision is to change the dialogue around death and become a movement for change, whereby both life and death should be discussed, celebrated and managed properly. Aura has similar features to other sites although it does also provide a Life Story feature which allows people to curate and highlight moments of their life. Other sites such as MuchLoved allow people to post and share funeral notices for themselves and RecordMeNow is a mobile app that offers a set of prompts to help people build a video library for their children, ranging from songs to encouragement to live a happy life.

One of the difficulties with those left-behind identities is that they cannot be erased and therefore the living may have ambivalent views about the relationships with the dead, post-death. Žižek (1999), in his deconstruction of *The Matrix,* suggests the possibility that the deletion of our digital identities could turn us into 'non-persons' – but perhaps a more accurate idea would be one of having left-behind identities rather than being seen as deletions. Thus, as we shift and move identities across online contexts, rather than deleting those that become superfluous, we tend to leave them behind and those who are left behind to deal with them may find them more troublesome than helpful.

Spiritual Beings

A different kind of relationship is that with spiritual beings. For example, ancestors, ghosts and some demons and deities are former humans and these spirits share the world with humans in parallel dimensions. Ghosts may also be seen as the souls of the restless dead who have been sent to warn the living, such as Marley's ghost in Charles Dickens' *A Christmas Carol* (1843), who warns Ebenezer Scrooge of the need to stop being self-ish and treating people badly in order to avoid rotting in hell. Shakespeare uses ghosts as warnings in *Richard III* and *Hamlet*. In *The Tragedy of Richard III* (Shakespeare, 1633), Richard is asleep in his tent before the Battle of Bosworth Field. He is visited by the spirits of his victims, one after another, each one predicting his death and telling him to 'Despair and die' (Act 5 Sc 3: 126). In *The Tragedy of Hamlet, Prince of Denmark* (Shakespeare, 1609), the old King, having been killed by his brother, appears on the battlements at dawn asking that his son revenge his death. In more recent stories such as *The Shack* (Young, 2007) ghosts or visions help the family and police to discover the identity of the murdered, and in *The Lovely Bones* (Sebold, 2003) Susie Salmon, a 14-year-old girl who is raped and murdered in the first chapter, narrates the novel from Heaven and could be perhaps classed as a narrating angel.

Walter sees the prevalence of angels in online death communication as 'a new religious discourse' (2019: 386) which is used as a coping mechanism or even a metaphor, even though spiritual beings are invariably still seen as religious figures. In particular, he argues that angels are seen to have agency– continuing their earthly activities in heaven whilst at the same time looking after those on earth who still need their care and guidance. It is evident from research in this area, for example Walter (2016), that many people see their dead relatives in cyberspace as angels. The idea here is that humans become angels, rather than the theological stance towards angels of being separately created celestial beings. This complex relationship with the dead in cyberspace, particularly the idea of then becoming angels, is discussed in Chapters 2 and 8.

What is clear is that exploration of the afterlife introduces questions about what it means to be human. Harari, in *Homo Deus* (2015), explores the projects that will shape the 21st century. Much of what he argues is disturbing, such as the idea that human nature will be transformed in the 21st century, as intelligence becomes uncoupled from consciousness. Google and Amazon, among others, can process our behaviour to know what we want before we know it ourselves, and as Harari points out, somewhat worryingly, governments find it almost impossible to keep up with the pace of technological change. Harari challenges us to consider whether indeed there is a next stage of evolution and asks fundamental questions such as where do we go from here? Perhaps what it means to be human is bound up with the understanding of consciousness or the idea of having a soul. This is something Harari explores but he seems to mix up religion and values by using the label 'religion' for any system that organizes people, such as humanism, liberalism and communism. Furthermore, there is also confusion about the relationship between the afterlife and the idea of heaven, with the assumption that they are necessarily the same thing. In fact, in many religions this is not the case since movement into heaven is expected to occur, certainly in the Christian tradition in the end times, and thus afterlife is more akin to a place of resting or sleeping immediately after death.

CONCLUSION

Although digital afterlife has several components and has been explored recently in depth by some authors, this is not the case with the spiritual realm, which is broadly understood to be both a liminal space and also a space that appears to sit alongside the physical world. Despite digital media appearing to be changing understandings of grief and the afterlife, there has been no sustained research that has examined its effects on faith, religion or comprehension of the spiritual realm. Further, understandings of the spiritual realm tend to vary across religions and belief systems, as will be explored next in Chapter 2.

CHAPTER 2

Religion and Afterlife

INTRODUCTION

This chapter begins by exploring the meaning of death and presents perspectives on death and the afterlife in relation to six different belief systems, namely Buddhism, Christianity, Hinduism, Humanism, Islam, Judaism and Sikhism. It draws on interview and email conversations from those of particular faiths in order to compare diverse and overlapping perspectives on the afterlife. The chapter presents a very broad overview of different religions and therefore does not cover the diverse differences within a given religion in detail. The last section of this chapter examines post-death practices and current trends in death merchandise and grave goods.

MEANINGS

The meaning of death differs across religions and this therefore has an impact on how death and the afterlife are perceived in relation to the digital. Death is not actually something we experience since, as Wittgenstein noted, 'death is not an event in life: we do not live to experience death' (1961, 6.43 II). The earliest responses to death and funeral rituals are found in Persian, Judaism and Greco-Roman culture exemplified in the Iliad (Homer et al., 762 BC/1998). However, in historical terms, homo sapiens have been undertaking funeral practices and rites for many thousands of years, and there is evidence that Neanderthals deliberately buried their dead in particular places ('cemeteries') (Than, 2013) and may have laid flowers on the bodies before covering them. Cremation as a funeral practice has been practised at least since 3000 BCE in the British Isles alone,

DOI: 10.1201/9781003098256-2

along with de-fleshing and burying bones in familial groups in long barrows which were practised for at least 1,000 years previously.

The Hebrew Bible and The Tanakh contain many examples of private and public mourning, such as Abraham weeping for Sarah (Genesis 23: 2) and all of Jerusalem mourning for Josiah (2 Chron 35). The theologian Jürgen Moltmann (1996: 119) argued 'the greater the love, the deeper the grief' and in biblical literature weeping is associated with mourning. The examples of mourning and lament seen in early practices illustrate the shifts that have now occurred away from the importance of grief customs and rituals. Today, in the west, mourning tends to be an isolated and isolating existence. Models of stages of death as suggested by Kubler-Ross (1969) and stages of grief and mourning (Murray-Parkes, 1971) have already been taken over by theories such as continuing bonds. Throughout most of the 20th century, it was expected that grieving involved a process of letting go, breaking bonds with the deceased. The continuing bonds model suggests that the relationship with the deceased changes, rather than breaks (Klass et al, 1996). Whilst the theory of continuing bonds is useful, it is still in many ways just a theory, and it is important not to normalize grief and mourning through theories and particular linguistic devices as grief is both personal and contextual. Recent research and literature on grief argue for the need to ensure grief is contextualized and points out that social media can seek to protect the living from the dead in cyberspace in ways that are not always appropriate so that protection is imposed upon them (for example, Giaxoglou, 2020; O'Connor and Kasket, 2021). However, in some social media spaces, grief and mourning do reflect the idea of the sequestration of death. The sequestration of death is the idea that the organization and the engagement of death practices have become separate from society (Mellor and Shilling, 1993) so that we have become a 'death denying society' (Becker, 1973). Such sequestration seems to result in a compression of the grieving process instead of a recognition of the need for a long grieving process whereby acceptance is not just about the acceptance of death but about the acceptance of a different kind of life. The theory of sequestration has been critiqued. For example, Maddrell (2016) argues for the importance of spatial dimensions of grief, mourning and remembrance which are threefold: physical spaces, embodied-psychological spaces and the virtual spaces of digital technology. Walter (2019) suggests the need to consider body, spirit and mourners together, as what he terms 'the pervasive dead', which he suggests removes the idea of the dead's separation from living society and instead 'bonds continue, the online dead can appear at any time, human remains sustain the everyday environment, and the dead become angels caring for the living' (p. 389). However, I suggest that sequestration does still occur in the UK society, although it is perhaps less prevalent with the emergence of the digital, which appears to reduce such sequestration.

RELIGIOUS PERSPECTIVES ON LIFE AFTER DEATH

Although there are over 4,000 religions in the world, 6 religious stances towards death and the afterlife are presented here, along with views from people of faith. Each section provides a basic overview of the faith doctrine and a stance towards the afterlife.

Buddhism

Buddhism teaches that an individual is but a transient combination of the five aggregates (*skandhas*): matter, sensation, perception, predisposition and consciousness and so has no permanent soul. The human being is an assemblage of various elements, both physical and psychical, and none of these individual aspects of a whole person can be isolated as the essential self nor can the sum of them all constitute the self. Since a human is composed of so many elements that are always in a state of flux, it is impossible to suggest that an individual could retain the same soul-self for eternity. There are many schools of historical Buddhism: Hinayana, Mahayana, Tantric and Pure Land, and it is difficult to find consensus among them concerning the afterlife. Tibetan Buddhism's *Book of the Dead* provides an important source for an understanding of their concept of the afterlife journey of the soul. A lama (priest) sits at the side of the deceased and recites texts from the *Book*, a ritual which is thought to revive the *bla,* the life force within the body, and give it the power to embark upon a 49-day journey through the intermediate stage between death and rebirth. Across the different forms of Buddhism, the notion of soul persistence and reincarnation is very strong in some and less so in others. As Liz, an academic colleague, explained:

> My Buddhism readings have led me to understand that sense of self is an illusion. But then I follow Zen teachings, and not all Buddhism traditions are the same. The notion of soul persistence and reincarnation is very strong in some, depending on the older religious traditions that Buddhism was overlain on. Also, in Buddhism generally, the idea of being reincarnated (samsara) until you reach a state of enlightenment and can finally stop having to

go round and round periods of existence (reaching nirvana) is a very old tradition. Only then do you free yourself from suffering – well, *dukkha*, which translates more like a wheel out of kilter than what we understand to be suffering. So I guess my short answer would be, it depends on which Buddhism tradition you are referring to, and there is a huge literature arguing whether reaching nirvana leaves a 'residue' or whether it is total annihilation.

In Buddhism generally, the idea of being reincarnated (*samsara*) until you reach a state of enlightenment is at the heart of many teachings.

Christianity

Whilst there are varying beliefs about what happens post-death, one straightforward argument has been presented by McDannell and Lang (2001), who suggest that either the entire person will be resurrected or the soul goes to heaven and the body is left on earth. Gooder (2011: 80), however, suggests that there is a need to rethink the idea of the soul:

> . . . the Hebrew often translated as 'soul' (e.g. Ps 62.1 'For God alone my *soul* waits in silence') is often elsewhere translated as 'life' (e.g. Ps 59.3, 'Even now they lie in wait for my *life*'). The two are almost interchangeable in the Hebrew Bible and the words in italics translate the same Hebrew word, *nepesh*.

Gooder suggests that it is therefore misleading to translate *nepesh* as a soul at all since this suggests a separation of the soul from the mind and the body. Before a belief in resurrection emerged, life after death in the Hebrew Bible was not about resurrection but about the continuation of the family line, which was why the birth of a son was so important then. Life and death were highly relational events in traditional Israel, bound up in the community and with a focus on 'biological immortality', the importance of transmitting our genes via our descendants. Recent studies, such as Cook (2007), make cross-cultural comparisons which explain the vital role of the dead in maintaining ties through lineage, and tenure of and burial on ancestral land.

An area that still remains deeply troublesome is where the dead go after life. Gooder suggests that Sheol is the fate of everyone who dies, good and bad – apart from Enoch and Elijah, two Old Testament characters who just went straight to heaven (Gen 5: 24; 2 Kings 2: 11). Sheol is seen as downward and the opposite of heaven but not the same as hell. Whilst

Gooder suggests Sheol is for the good and the bad, Cook (2007) argues Sheol is for those whose lives have gone wrong, not the godly. Further, he suggests that the practice of relatives leaving lamps for the dead may be due to the perception of the darkness of Sheol (Job 10:21–22). Suriano (2016: 3) suggests that the paradox of Sheol is because it is seen as a dismal place for all humanity, rather than some form of peaceful reunion for the dead. Cook (2007), in a rather similar way to Gooder, suggests that Sheol may have tiered strata so that the vile is in the deep abyss (Ezekiel 32: 23). In an interview with an ordained participant, Anna, who has studied death in Victorian times, they explained their stance.

> I believe that beyond death there is life and there is God. Anything beyond that is speculation. There are things that I would want to speculate on based in my faith and my theological distillation of my faith. I would probably say that I don't have a Calvinist view that how you are at death is how you're set for the whole of eternity, because I see in creation, which I believe to be God breathed, I see development, I see progress, I see growth. I cannot really conceive of a God who would not allow for post mortem growth, development, spiritual progress, however you want to describe it.
>
> *Maggi: Is that the idea of stages of heaven?*
>
> I don't quite go with that. I don't like the whole stratified divine comedy or even the spiritualism that I've done quite a lot of research into, which does have a stratified heaven and you rise up through the spheres, dependent on your spiritual progress. I don't quite like that, but we're drawn to God in life. We grope for the spiritual, we grope for the divine in a whole variety of different ways. I'd like to think that that wouldn't stop.

However, debate remains about the extent of clear reference to life after death. Gooder suggests the most notable is that in Daniel 12: 1–3, 13:

> At that time Michael, the great prince who protects your people, will arise. There will be a time of distress such as has not happened from the beginning of nations until then. But at that time your people – everyone whose name is found written in the book – will be delivered. Multitudes who sleep in the dust of the earth will awake: some to everlasting life, others to shame and everlasting contempt. Those who are wise will shine like the brightness of the heavens, and those who lead many to righteousness, like the stars

for ever and ever ... As for you, go your way till the end. You will rest, and then at the end of the days you will rise to receive your allotted inheritance.

What is interesting about this passage is that those who rise are sleeping in the dust of the earth – which implies resurrection, for 'many', but not all. Gooder (2011: 87) offers an insightful view of Hebrew and Greek thought on this: there is no evidence to support that in Hebrew thought no person could survive outside the body, whereas according to Greek thought, the body could be split for the soul or spirit. What she does suggest is that an embodied resurrection fits well with notions of Jewish humanity and the idea of living well. Images of resurrection, however, are unclear. For example, there is the possibility of transformation in stars or angels, as implied above in Daniel 12. There is the suggestion of a dead person rising in Isaiah 26 and the possibility of waking from sleep, also in Daniel 12.

Cook suggests that there is more in the Old Testament about the afterlife than most people realize. He argues:

The soul's continuation despite death was assumed in Israelite culture, as it was in Israel's ancient Near Eastern milieu. Among the ample textual evidence for post-mortem survival in the Hebrew scriptures are the following passages, which are particularly clear about the matter: 1 Samuel 28; Isaiah 8:19, 10:18, 14:9–10; Ezekiel 32:21; and Genesis 35:18.

(Cook, 2007: 668)

However, other theologians would argue against Cooke's suggestions since, despite Jesus' rebuke to the Sadducees in the Gospel of Mark chapter 12, the reasons they did not accept resurrection was based on its absence from the Torah, and thus they would not accept that afterlife was mentioned in Genesis 35. Moving away from the Old Testament, the focus in the New Testament is very much on bodily resurrection based on faith. What is clear from Paul's stance in 1 Cor 15 is that there will be a difference between pre- and post-resurrection bodies. Thus, Paul sees us as having bodily resurrection in a recreated world.

Hinduism

In Hinduism, death is not seen as the end but more as a transition, and a person's immortality has a shared identity with Brahman, the Supreme

Being. Hindus believe that reincarnation lasts until *moksha* occurs, when the soul is liberated. Thus, spiritual essence is the divine part of a living being, the *atman*, which is eternal and seeks to be united with the Universal Soul, or the Brahman. The belief stems from writings known as Upanishads that set forth the twin doctrines of *samsara* (rebirth) and *karma* (the cause-and-effect actions of an individual during his or her life). Hindu cosmology, based on the Vedic literature depicts three *lokas*, or realms: heaven, earth, and a netherworld. *Loka* does not just mean world or place but rather a religious spatiality with soteriological value. However, in the Hindu text *Brahmanda Purana*, there are 14 additional worlds (*lokas*) in which varying degrees of suffering or bliss await the soul between physical existences. Seven of these heavens or hells rise above the earth and seven descend below. According to the 9th-century Hindu teacher Sankara, the eventual goal of the soul's odyssey was *moksha,* a complete liberation from *samsara,* the cycle of death and rebirth, which would lead to nirvana, the ultimate union with the divine Brahma. As one Hindu, Raj, a business consultant, explained:

> My views about afterlife are that the soul leaves your body and after some time you are reborn according to Karma . . . According to Hindu religion your soul travels from body to body until you get Moksha and I believe in it. You only receive Moksha according to your Karma. As my personal experience, I believe that your soul lingers around your loved ones until its ready to leave and take another form.

However, it is important to note that Hindus in different states in India have different funeral types. Similarly, there are different beliefs about the soul so that some believe that the soul stays for 4 days after death and others 13. Hindus believe that an individual has a direct influence on his or her karma process in the material world which determines the form of his or her next earthly incarnation.

Humanism

Humanists have no belief in an afterlife, and so they focus on seeking happiness in this life. They rely on science for the answers to questions such as the creation and base their moral and ethical decision-making on reason, empathy and compassion for others. Humanists argue that human values make sense only in the context of human life and therefore existence after death is not part of the humanist value system. The current world is the focus of their ethical concerns and aspirations and therefore the values they hold are placed in the context of the here and now.

Islam

Muslims see death both as the return of the soul to Allah and as the transition into the afterlife. Between death and judgement, the soul remains in a waiting state. The concept of a soul in Islam sees a human as a being of spirit and body. The Prophet Muhammed PBUH (570 –632 CE) regarded the soul as the essential self of a human being, but he, reflecting the ancient Judeo-Christian tradition, also considered the physical body as a requirement for life after death. The way one lives on earth affects the afterlife, and there are promises of a paradise or warnings of a place of torment. In death, the body remains in the ground while the soul is in the interspace or Barzakh between the two worlds which are still connected, and so the bliss or punishment happens to both of them. The belief is that when Allah desires bliss or punishment for the soul, he connects it to the body. However, this is dependent on earthly actions and on the will of Allah. The Prophet PBUH speaks of the Last Judgement, after which there will be a resurrection of the dead which will bring everlasting bliss to the righteous and torments to the wicked. Judgement is individual so that it is not possible to help a family member or friend. Faith in an afterlife is based upon the belief in the oneness of God and the belief in a day of resurrection and judgement for all, regardless of religious beliefs. At that time, the spirit will be judged, based upon its deeds in life, and allowed either to enter into Paradise and be with God or be thrown into the fire for a period of purgation or condemned to everlasting punishment in the Fire. Most Muslims believe that non-Muslims can reach Paradise only after a period of purgation. The Islamic religion views death as a transition to another state of existence called the afterlife. Where you go in the afterlife depends on how well you followed Islamic religious codes during your life. In the Islamic tradition, close members of the family perform the ritual washing of the body. Once the body is prepared for burial, mourners are welcomed to view the body of the deceased as a reminder that every soul shall taste death.

Judaism

In Judaism, the belief is that life does not begin with birth nor does it end with death. This is articulated in the book of Ecclesiastes 12 v 7, 'And the dust returns to the earth as it was, and the spirit returns to G–d, who gave it'. Central to the Jewish faith is the belief in *techiat ha-meitim* (resurrection of the dead) and that the soul will be restored to a rebuilt and revitalized body. For Jews, the soul is the higher, more spiritual incarnation of the self, but they maintain high respect for the body since it is seen as the mechanism which enables all the soul's accomplishments during life.

Jews do believe in an afterlife but the formulation of this varies; there is a sense that the soul journeys, passing such things as *Dumah*, an angel of the graveyard, and Satan as the angel of death on the way towards *Gan Eden* – heavenly respite, or paradise. Jewish eschatology combines the Resurrection with the Last Judgement. Resurrection is referred to in the book of Ezekiel, chapter 37, as only being for Jews, but later resurrection underwent a change and was made part of the Day of Judgement on the basis of Daniel 11v 2. The resurrection includes both the righteous who awake to everlasting life and the wicked who face shame and everlasting horror. Those who believe in God's judgement believe that the decision about whether they should be rewarded or punished is based on how well they have followed the mitzvot – the Jewish laws and commandments. However, conservative Jews tend not to believe in a resurrection and instead argue that concepts such as heaven and resurrection are metaphors.

Sikhism

Sikhs believe that upon death one merges back into the universal nature. Sikhs do not believe in heaven or hell. Heaven can be experienced by being in tune with God while still alive. The bodies of the deceased will later be cremated, but their souls will live on.

Sikh tradition teaches. Sikh scriptures do not dwell on what happens after death. Instead, the faith focuses on earthly duties, such as honouring God, performing charity and promoting justice. Sikhs believe in karma or 'intentional action'. As one Sikh participant, Jaswinder, explained:

> I believe your soul joins the Almighty when you die. In practice in my religion there is the vague notion of a 'good place' where the Almighty welcomes the souls of those who have followed the principles of the religion, undertaken *seva* and an honest living whilst those who have not are condemned to the cycle of rebirth.

Through good actions and by living a good life and keeping God in their minds, Sikhs hope to achieve good merit and avoid punishment. In Sikh theology, living morally prepares the soul to receive God's grace. For those fortunate enough to escape rebirth, the ultimate destination is a return to the divine soul from which all beings emanate. The tradition also teaches us that death is something worth celebrating because the souls are going to be one with God. Table 2.1 presents a summary of afterlife belief along with perspectives garnered from those interviews whose date is presented in depth in Chapter 9.

PRACTICES

Practices around the management of the dead vary across cultures, religions and countries. This section explores the diverse practices across the different belief systems.

Funeral Practices and Rituals

The Buddhist rituals are some of the most comprehensive post-death rituals but vary according to the type of Buddhism. For example, Buddhists believe that chanting texts from Buddhism will generate merit that can be transferred to the deceased and help them in their rebirth. During the Bardo, the 49 days between death and when rebirth is thought to occur, relatives read texts specific to any practices the deceased favoured. The readings help the deceased in their journey to rebirth. In many Buddhist temples in Japan, there are small statues that are a combination of a Buddhist monk and a young child. The statues are *mizuko jizo* and they represent the spirits of stillborn or aborted foetuses (Klass and Heath, 1996–97). Theravada Buddhists believe they will gain favour for the dead by offering the monks white cloth to be used in the creation of robes. Chinese and Laotian Buddhists celebrate Ghost Month, which is believed to be when the gates of hell open and hungry ghosts are thought to walk the earth in search of food and gifts. Therefore, food, incense, paper money and other gifts are offered to the deceased spirits to ensure good merit.

Christian practices vary across different traditions with burial or cremation being acceptable. Black is often worn but increasingly the funeral is seen as a place to celebrate life and often bright colours are worn. In Christian traditions, there are different views about praying for the dead, for example, Catholics will pray for the souls of the dead whilst Protestants tend to pray for the families of the dead. What is not common across traditions is embalming. Embalming is used by Christians but is not a practice or ritual used in Buddhism, Hinduism or Islam. Its use stemmed from a medieval belief that the preservation of the body indicated a sign of favour

TABLE 2.1   Religion and Afterlife Beliefs

Religion	Afterlife belief	Personal stance towards digital afterlife from research participants
Buddhism	Soul persistence and the idea of being reincarnated (samsara) until you reach a state of enlightenment is at the heart of many teachings	It is an interesting idea to help those grieving but will not facilitate reincarnation
Christianity	Those who believe in Jesus will be reunited when Christ returns	Acceptance that this may be useful for some people to cope with grief but puzzled as to the value given afterlife will be with God in heaven
Hinduism	The soul leaves the body and is reborn. An individual has a direct influence on his or her karma process in the material world which determines the form of his or her next earthly incarnation	Concerned it would affect the loved ones, be upsetting for those left behind as well as affecting karma
Humanism	There is no afterlife	It is an interesting idea to remember someone, but as there is no afterlife, it was perceived as being somewhat futile
Islam	Death is a transition to the afterlife but where you go in the afterlife depends on how well you followed Islamic religious codes during your life	Since death is a transition into the afterlife, a digital afterlife may affect judgement
Judaism	The afterlife is known as the world to come and there is a belief in the journey of the soul after death for some Jews; for others, the notion of the soul and the idea of a resurrection is seen as purely metaphorical	There is nothing wrong with leaving behind a digital afterlife for the family as a form of remembrance or creating a digital avatar.
Sikhism	The soul is united with the almighty depending on its deeds done in this life	It may be helpful as long as it is a true reflection of self rather than an idealized copy

from God. However, Cann (2018) suggests the development and continuation of embalming in the US is not just a result of religious beliefs. Cann argues:

> that the privatization of American industry and American denominationalism went hand in hand to create a ripe consumer culture in which funeral directors were able to effectively

create a cultural rhetoric surrounding death and grief, in which embalming became a symbol of American progress, modernity, success, and the manipulation of time (corpses that can wait on the living vs. the other way around), bodies, and space.

(CANN, 2018: 385)

Thus, embalming has been affected by the medicalization of death, the growth of funeral homes as an industry and the professionalization of dying and death. However, it is important to note that Christian practices at death are not universal and are culturally determined so, for example, Christians in Africa will follow traditional practices and Orthodox Christians do not cremate.

Hindus are cremated since fire burns the ties to the physical world and ensures that the soul makes the necessary transition. Death rituals are important in demonstrating the indebtedness to ancestors and the repaying of debts to the said ancestors. In Hinduism, the cremation of the deceased marks the beginning of the mourning period, which lasts for 13 days, and the family of the deceased will stay at home and receive visitors. Throughout the mourning period, the rite of *preta*-karma is practised, a ritual that helps the deceased person's soul move from spirit form to its new body in the cycle of reincarnation. The rituals such as placing water and a tulsi leaf from the river Ganges are designed to wash away the sins of the departed. One year after the death, the family will observe a memorial event, a *sraddha*, which pays homage to the deceased.

Humanists do not have any particular rituals as they do not believe in an afterlife. For Sikhs, the funeral is known as *Antam Sanskaar*, 'the last rite of passage'; thus the funeral focuses on the celebration that the soul has an opportunity to rejoin God. Sikhism does not subscribe to any mourning periods or mourning rituals after the death of a loved one.

According to the spiritual traditions of Judaism articulated in the Talmud and Kabbalah, the soul does not completely leave this world until after the burial. Those present at the time of death recite the blessing: *Baruch Dayan Ha'emet* ('Blessed be the True Judge'), and after the death is confirmed, the eyes and mouth of the deceased are closed and a sheet or other cover is drawn over the person's face. The body of the deceased is then placed on the floor, and while the body is lowered to the floor, forgiveness should be asked of the deceased and then candles are lit near the deceased's head. There should always be someone with the body until the funeral and they should recite prayers or psalms as this brings comfort to the soul of the deceased. The body is not embalmed or cremated and the burial should take place as soon as possible, preferably on the day of death.

Muslims are always buried, as the grave is seen as the first phase of the afterlife and the fate of the deceased is decided upon during the first night in the grave. The deceased lives on in a transitional world. After death, it is believed that the deceased encounter two angels who visit them to test them on their belief in the prophets, God and religion. If they answer correctly, the grave is illuminated and expanded until judgement day. The funeral rituals of good deeds undertaken by those left behind are designed to help the deceased with the test so that Allah will be merciful.

CONSUMER DEATH GOODS

Death goods are artefacts such as coffins, urns and plaques that are sold by funeral companies. In the 2000s, all of these can be bought on Amazon and in the US from superstores such as Walmart. These are traditional goods but it is also possible to buy more unusual goods such as bullets in which to place the ashes. Cann (2018) points out that consumer choices are not just based on cost but on beliefs about the afterlife. This would include issues such as what might help the deceased to make the transition to the afterlife and the value, or not of, embalming, burial or cremation. Recent innovations in death goods include designer coffins – that begin to look more like art installations and are available at Crazy coffins in the UK, illustrated in Figure 2.1; and ashes are put into bullets so they can be fired from your favourite weapon, illustrated in Figure 2.2.

And Vinyly – https://www.andvinyly.com/

This is a service that uses a teaspoonful of someone's ashes to create a playable vinyl record, illustrated in Figure 2.3. The company's tag line is that you can 'Live on from beyond the groove'. The cost of the record ranges between £1000 and £3000, depending on the specification and the quantity.

EverWith – https://www.everwith.co.uk/

This company makes handcrafted memorial jewellery including rings, pendants, earrings and bracelets. The cremation ashes are set in a choice of resin colours mounted on the jewellery. The company states that no ashes are discarded and all unused ashes are returned to the sender.

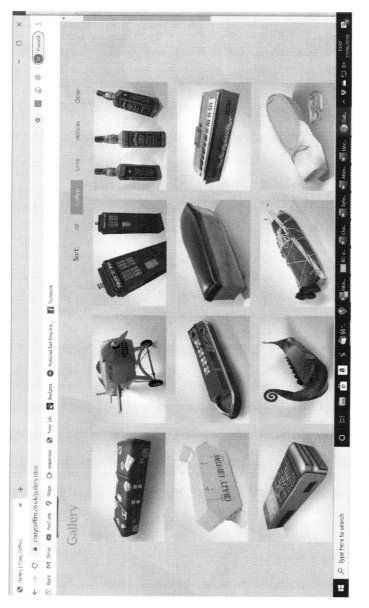

FIGURE 2.1   Crazy coffins (https://www.crazycoffins.co.uk/index.html).

This is the perfect addition to a memorial service for veterans, police officers, hunters or marksmen. You get to pick from a variety of rounds and shells to suit the shooter's favorite pistol, rifle, or shotgun. Simply send us the ashes, and we'll create the rounds by an FBI approved re-loader. You can expect professional service from an experienced, federally approved loader. These Memorial Shots© create a special way for you to remember your game hunter. The rounds are perfect for a 21 or another gun salute to your special loved one. At In the Light Urns, we design custom cremation urns, jewelry urns, and other keepsake pieces to hold your loved ones' ashes. For more information or to place an order, call 800-757-3488 between 8:30 AM PST to 6:00 PST 5 days a week. You can also contact us through our LIVE SUPPORT feature on our website at (https://goo.gl/UX5pJT).

#ammo #gun #death #miami #florida #seattle #portland #lasvegas #nevada

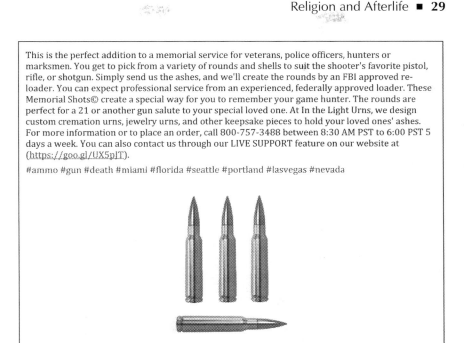

FIGURE 2.2 Shooting your loved one's ashes (https://www.inthelighturns.com/).

Space Burials

Whilst many of the artefacts are designed to be worn or kept, it is also possible to have a 'Space burial', illustrated in Figure 2.4. Instead of being shot from a gun, the ashes of the deceased are launched into space. This service is provided by the company Celestis. The company's official homepage states that the ritual consists of 'launching a symbolic portion of cremated remains into near-space, Earth orbit, to the lunar surface or even beyond' (Celestis n.d.). It seems quite an odd choice since the living will watch it disappear but the dead will probably have no knowledge of this experience.

Grave Goods

Traditionally grave goods were the items buried along with the body which included personal possessions. There were also artefacts or food and drink, designed to ease the deceased's journey into the afterlife, provide them with what was needed in the afterlife, or in some cases they were offerings to the gods. Perhaps the most well-known grave goods are those found in Egyptian tombs which included not only gold and silver objects but also board games, chairs and clothing. Most grave goods recovered by

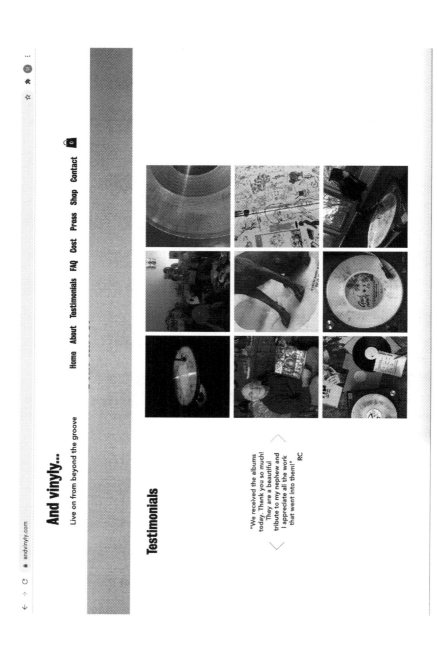

FIGURE 2.3 And Vinyly.

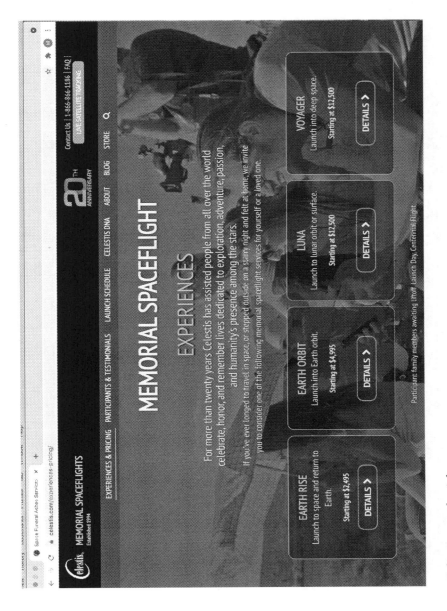

FIGURE 2.4 Celestis space burials.

TABLE 2.2 Funeral Practices, Rituals and Grave Goods

	Funeral practices	Rituals	Grave goods
Buddhism	In general, there is a funeral service with an altar to the deceased person. Prayers and meditation may take place and the body is cremated after the service. Buddhists believe that cremation is an important ceremony for releasing the soul from its physical form	Many Buddhists believe that chanting texts from Buddhism will generate merit – merit that can be transferred to the deceased and help them in their rebirth. Other practices include paper goods offered in Ghost Month and annual offerings of food	In Buddhism, votive offerings such as the construction of stupas was a prevalent and holy practice in ancient India. Today, votive offerings, if undertaken at all, usually take the form of a small clay or terracotta tablet bearing Buddhist images
Christianity	Burial or cremation is permitted. In general, there is a church service followed by cremation or burial. Christians believe that the soul leaves the body at death and therefore cremation or burial is acceptable	Lighting candles to remember the dead. In some forms, praying for the souls of the dead	Christians do not tend to leave grave goods. Flowers are left on the grave and a headstone is added later at the graveyard. Ashes are scattered or kept in an urn after cremation. The Church of England has strict rules about burial plots and what can be included on a headstone
Hinduism	There is no burial. The bodies are burned. The body remains at the home until it is cremated, which is usually within 24 hours after death. The ashes are typically scattered at a sacred body of water or at some other place of importance to the deceased	Eck (1998) describes *darsana*, the interaction between a person and a Hindu god. The connection happens when the god and the person recognize each other.	After the funeral, rice balls are offered by the sons of the deceased to the departed soul, to help it to construct a body for its existence in the world of the ancestors. This occurs for ten days after which the ghost body will be ready. It is followed by another rite known as *sapindakarana*, which facilitates the entry of the soul into the world of ancestors and its continuation from there on

(Continued)

TABLE 2.2 (CONTINUED) Funeral Practices, Rituals and Grave Goods

	Funeral practices	Rituals	Grave goods
Humanism	Humanists believe that science provides the only reliable source of knowledge about this universe and that there is no afterlife. Humanists can be buried or cremated, and funeral services are conducted by a humanist celebrant, family member or close friend and do not include hymns or prayers	No rituals as they do not believe that soul exists after life. An example of this is a *direct cremation* where the cremation takes place completely separately from the personal farewell. For example, see https://www.purecremation.co.uk/	Humanists do not leave grave goods as they do not believe in an afterlife
Islam	The deceased are buried within 24 hours. Bodies are not embalmed as this is seen as disrespecting the body. There is no viewing, wake or visitation. Since Muslims believe there will be a physical resurrection of the body on Judgement Day, the faith prohibits cremation. Similarly, autopsies are strongly discouraged since they delay burial and are considered a desecration of the body	Sometimes a family a member may make a hajj (a pilgrimage) on behalf of the dead person	After the burial, Muslims pray for the forgiveness of the dead. This is the last formal collective prayer for the dead. In some cultures, e.g. in South-East Asia, the relatives scatter flowers and perfumed rose water upon the grave before departing

(Continued)

TABLE 2.2 (CONTINUED) Funeral Practices, Rituals and Grave Goods

	Funeral practices	Rituals	Grave goods
Judaism	The body is not embalmed or cremated and the burial should take place as soon as possible, preferably on the day of death. There should always be someone with the body until the funeral and they should recite prayers or psalms as this brings comfort to the soul of the deceased	The first seven days after the funeral is known as 'shiva' and the deceased's family stay at home and receive guests, pray and reflect upon their loss. Second mourning lasts 30 days after the funeral. On the anniversary of the death, a candle is lit and left to burn for 24 hours. This is known as yahrzeit	Grave goods are not used and flowers at the grave are strongly discouraged; however, there is a Jewish practice of placing a pebble on a grave to remind the dead that the living have not forgotten them
Sikhism	Cremation is normal, and there is a religious service at the Gurdwara after the cremation. In other parts of the world, this is done with outdoor funeral pyres, but in the UK, this is restricted to crematoria. Burial or any other means for disposing of the body are acceptable if the circumstances do not allow for cremation	On the first anniversary of the death of a loved one, the family gathers together. There is a celebration of life, time for prayers and a large meal	Sikhism teaches that the goal is to escape from the cycle of death and rebirth but does not believe they liberate themselves; only God's grace offers freedom from rebirth. Grave goods and votive offerings are not used by Sikhs

archaeologists consist of inorganic objects such as pottery and stone and metal tools. The material objects discovered in Israelite tombs show they provided their dead with food and drink. This is also an indication of an increasingly affluent society, even as recently as the 19th-century common (unmarked) graves were used to bury the poor.

Today grave goods are still buried with the dead and hold different significance depending on the religion. For example, the Bantu-speaking people of south-western Zimbabwe (Ndebele) kill an animal after burial in order to provide food for the journey ahead. In China, many people believe in leaving artefacts such as paper money or phones at the graveside for ancestors to use in the afterlife. The use of grave goods and funeral practice is linked to the ways in which different religions perceive the afterlife. For example, Cook (2007) explains that for the Akamba, in Kenya, the physical burial of the corpse with their ancestors symbolizes but also ensures a warm welcome within the ranks of dead relations. However, if someone dies far away and the remains are unable to be taken home, the Luo of Kenya people bring handfuls of dirt home to be placed in the graveyard where they feel that the body really should be lying, which would seem to equate with the western practice of cenotaphization. Table 2.2 provides an overview of some of the rituals and grave goods by belief system.

However, it is important to note that there are increasing discussions now about what might be considered to be a grave but also the whole notion of grave goods. For example, McKinley (1994) distinguishes between pyre goods (items placed on cremation pyres and thus burned, distorted, fragmented/partial) and grave goods (as in whole unburned objects). Further, there are increasing concerns about what is left at gravesides, as well as the pollutants caused by the burning of toxic materials with the dead.

CONCLUSION

It is evident that religious practices relating to death and the afterlife still vary widely in terms of earthly death practice. However, as can be seen from this chapter, as yet there is relatively little digital engagement that is directly guided by or prompted by religions. In the main, most religions do not teach or guide digital practices, apart from suggestions about what is seen as unhelpful or unacceptable. The difficulty here is that even though death does largely remain, for many, a religious event, the fact that the digital is largely left to one side has resulted in little understanding of the kind of impact the digital is having on the bereaved. This will be discussed next in Chapter 3.

CHAPTER 3

Impact of the Digital on the Bereaved

INTRODUCTION

This chapter explores the increasingly important area of the impact of digital afterlife creation on those who are left behind. It begins by presenting models of grief, from the traditional models of the 1960–70s to up to date perspectives on grief in a digital world. In particular, it discusses recent models that offer some purchase on understanding bereavement on social networking sites. The chapter then examines the impact of the digital on the bereaved. It analyzes the unintended consequences that the array of digital practices may have on the bereaved and outlines some of the new concerns that are having an impact on those who mourn, such as bandwagon mourning, trolling and grief policing. The final section provides recommendations of ways in which the bereaved might be supported better in digital spaces.

MODELS OF GRIEF

There are a number of models of grief, ranging from stage models to circular models. Kubler Ross (1969) argued for 5 Stages of Death which may not always be experienced in the same order. So, whilst it is often seen as a staged model, it is, in fact, a circular model with reoccurring stages, of

- Denial – refusal to accept facts, information, reality
- Anger – with themselves, and/or with others, especially those close to them

DOI: 10.1201/9781003098256-3

- Bargaining – attempting to bargain with whatever God the grieving person believes in

- Depression – often the beginning of acceptance, feeling sad, regret and uncertainty

- Acceptance – the beginning of emotional detachment

Murray-Parkes (1971) suggested a Phases of Grief model. This model considers people's history, experiences to date and in particular the relationship with the deceased; phases will differ in each case according to the relationship with the deceased. These four phases are:

- Shock and numbness

- Yearning and searching

- Disorientation and disorganization

- Reorganization and resolution

It is widely recognized that grief stages apply not only to grief following a bereavement but also to many other types of loss and trauma, such as amputation and personal reactions to change or life events. These stages have been recognized, adopted and adapted universally in the past. However, more recently different theories of loss and mourning have been suggested which would seem to have greater resonance with the idea of digital immortality. These include the following.

Continuing Bonds

Throughout most of the 20th century, it was expected that grieving involved a process of letting go, breaking bonds with the deceased. The continuing bonds model was developed by Klass et al. (1996), who argued instead that in fact the relationship with the deceased changes, rather than breaks. In short, the relationship does not end. This model challenged the popular models of grief requiring the bereaved to detach from the deceased. Thus, the model suggests that grief is not a linear process to be worked through which is completed when someone 'moves on'. Instead, after death, the relationship is redefined so that some kind of attachment is perceived to be normal. The focus is on a changed relationship, a continuing bond with the deceased, rather than letting go and moving on. However, recently

Klass noted that maintaining a connection with the dead is a common aspect of bereavement in all current models of grief, which is reflected by the use of digital and social media for this purpose (Klass, 2018). A similar model was proposed by Tonkin (1996), that of Growing around grief. This model suggests that grief does not necessarily disappear over time, but instead of moving on from grief, people 'grow around it'.

Dual Process Grieving

This model of grieving (Stroebe and Schut, 1999) suggests that there are two categories of stressors affecting bereavement: loss orientation and restoration orientation. Loss orientation refers to the way the bereaved person processes the actual loss. In practice, this means that they oscillate between confronting and avoiding issues connected with grieving. The restoration orientation refers to the secondary consequences of bereavement such as dealing with the world without the deceased person in it. The model also suggests that it is important to take respite from dealing with both of these stressors in order to adapt and cope.

Orientation to Network Grief

This is a new model of grief that was developed as a result of a five-year mixed methods study. Brubaker, Hayes and Mazmanian (2019) sought to understand how people positioned themselves and evaluated their expression of networked grief. What emerged from this study was a new cyclical model of grief pertinent to the digital age. It comprises stances that are dynamic and continuous.

Encounter. Here an individual encounters another's expression of grief on a social networking site which prompts the dynamic process of orienting to networked grief.

Situate. Once an expression of grief is encountered, the person seeks to locate themselves and the other person in relationship to their understandings of the network structure. For example, whether they are a close friend, family or celebrity, this categorization affects how they evaluate the expression of grief in terms of emotional content and appropriateness.

Evaluate. The individual evaluates the expression of grief based on the understanding of the network and by the norms an individual has learned for appropriate behaviour both on Facebook and grief more broadly.

Reassess. Re-evaluation of how the individual and others are situated in the networks, such as realizing someone may have been closer to the deceased than they first appreciated or that Facebook is too trite a space for grief management.

Establish orientation. Individuals establish an orientation in relationship to the network so that when they have a sense of the norms and expression of grief, they come to understand their own position within it.

Act. Once an orientation to the network is established, individuals make decisions about current and future encounters as well as their relationship to the network over time.

Whilst these models can be related to the creation of digital legacies and memorialization, a number of concepts and practices around grieving and the digital have developed in recent years.

DIGITAL GRIEF CONCEPTS AND DIGITAL GRIEF PRACTICES

Digital grief concepts are concepts that have been developed in order to make sense of the ways in which digital technology is being harnessed to commemorate and memorialize the dead, and digital grief practices are practices that have become acceptable norms through media and digital media.

Digital persistence. Kasket (2019) argues that online persistence and the ongoing presence of the data of the dead online will lead to more of a globalized, secularized ancestor veneration culture and that it is important to recognize the ongoing persistence of the dead online on social media, LinkedIn, Amazon and YouTube.

The restless dead. Nansen et al. (2015) argue that forms of digital commemoration are resulting in cultural shifts towards a restless posthumous existence. Thus, there is a shift away from the idea of death as being sleep or rest (Hallam and Hockey, 2001), towards the restless dead as they materialize through social media and technical capabilities. Such media include living headstones, digitally augmented coffins and commemorative urns embodying the head of the deceased, and which are seen to interrupt the previous limitations of cemeteries, static headstones and biological death.

Media mourning. Media mourning is defined here as the idea that we are urged to mourn something that is not our grief through social media, such as the 2017 Manchester Arena bombing, or US school and college shootings, or to mourn our personal loss through social media in a highly public way.

Durable biography. Walter (1996) argues that the purpose of grieving is to construct a durable biography that allows survivors to continue to integrate the deceased person into their lives and to find a stable and secure place for them. In practice, this now tends to occur more often through digital memorialization and the use of death tech companies.

Virtual veneration. This is the process of memorializing people through avatars in online games and 3D virtual worlds. An example of this is ancestor veneration avatars (Bainbridge, 2013) that are used as a medium for memorializing the dead and exploiting their wisdom by experiencing a virtual world as they might have done so.

Digital will creation. It is increasingly common for people to create a digital will in which they indicate what is to be done with their digital legacy and assets, and includes passwords and security questions. Digital will creation is dealt with in more depth in Chapter 10.

Intentional or accidental immortality. Intentional forms of digital immortality creation include transferring all assets of digital and non-digital media in a digital legacy, adding to current media memories predeath, creating a digital immortal (avatar) predeath and creating a digital immortal that would learn post-death. Most of these forms are likely to have been discussed with loved ones predeath and therefore the intentions and use of them are likely to be clear. What is more of a concern is if they have been intentionally created but not shared predeath. Accidental immortality occurs when media are preserved without the deceased person releasing it predeath. In 2019, Perfect Choice Funerals in the UK undertook a survey and found that the majority of people over 50 are not aware of the memorialization feature of social media profiles. They asked 1,000 people aged 50 and above about their knowledge of social media in relation to mortality. Around 70% of respondents were not aware that social media accounts could be memorialized. Those who were aware of memorialization online were then asked whether they had personally memorialized someone's profile, and they found that 3% have actually done it themselves, despite 18% saying that they had seen memorialized profiles (Perfect Choice Funerals, 2020).

Deletionist or preservationist stance. Whilst some people value engaging with the internet and digital media when they are alive, they may believe that part of the leaving, grieving and forgetting process for those left behind mean deleting all digital media. For preservationists, it is important in terms of our ancestry to preserve everything from online spaces. Differences in familial stances of a change of view post-death are likely to cause conflict and difficulty for the living.

What has become apparent is that these concepts, practices and decisions are having an (often detrimental) impact on the bereaved, with unintended consequences.

UNINTENDED CONSEQUENCES

The growth and changing nature of the digital has resulted in shifting norms and behaviour in online spaces. These shifts have resulted in mourning becoming something that is mediated and managed in new and different social spaces than previously. Klastrup has suggested that the lack of shared norms on social media has resulted in the compartmentalization of death; she suggested:

> This compartmentalization of death is in part due to the fact that people allegedly no longer share a religious vocabulary (or belief) with which to talk about and deal with death, leaving the deceased uncertain on how to grieve properly.

> (KLASTRUP, 2015: 149)

Spaces for mourning have expanded and have become more complex, with the result that often norms are developed within social media spaces and that participants learn about mourning norms by being in such spaces. Wagner (2018) reviewed a range of studies exploring mourning online and highlights that norms are constantly changing and being renegotiated by users of social media. A distinction is also made between those who are mourning themselves, and reactions to the mourning of others, highlighting the need to consider the needs of both mourners, and the users of social media who are reacting to expressions of grief online. Further Cann (2014) notes that in some instances grief rituals online and posting about the deceased are taken away from families as the primary mourners. This is because other people usurp this traditional hierarchy as it is not recognized in social media spaces, and often this privilege and privacy is not protected by other people who knew the deceased. One participant I interviewed, Anna, a minister in the Church of England, explained an incident where a friend felt she was not able to deal with the messages posted by other people on her dead sister's Facebook page.

> She has said how she's really struggling to read some of the messages on social media, people's responses to her sister's death. It's an outpouring of love and affection and sadness at the loss of such a lovely, vibrant, supposedly healthy woman, but she's struggling because of the unmanageability of it. You can manage your engagement with other people's grief a little bit more, so when you meet somebody in the street and they say, I'm so sorry to hear about your sister's death, then you can physically and quite clearly respond to that as you want. You have the power to walk away from it or to steer the conversation a bit. Whereas when my friend clicks onto Facebook and she sees tags about her sister and reads things, then she's not able to manage it in the same way. She finds that really quite hard. So that's a real tangible live example of something that I know about, even in the last three weeks.

Although in the past, death and mourning have been seen as private, the digital has resulted in more open spaces. Yet, the hidden nature of some activities means that one of the challenges of the digital age is that it is difficult to realize the impact on the bereaved. Much of what takes place is hidden:

- Hidden in the digital: such as keeping mobile phone messages of the deceased or creating digital albums of music or photographs. O'Connor (2020) undertook research that explores survivors' creative, relational and situated accounts of their dead. Data are presented from four interviews with a bereaved young woman, Sarah, as she told the story of her late sister, Leah, and their relationship. O'Connor argues that for survivors, the ability of the dead to 'live on' is only realized to the extent that material is woven into posthumous accounts, crafted around grievers' ongoing lives and experiences.

- Hidden from others: such as the quiet and secret creation of predeath avatars. Those creating a representative interactive avatar predeath design an avatar which is able to conduct a limited conversation with others but has a very limited capability to learn, grow, act on and influence the wider world around it (and hence could be considered a virtual humanoid in the typology identified by Burden and Savin-Baden, 2019). An example of this is Replika https://replika.ai/. Replika was founded by Eugenia Kuyda to enable people to create a personal artificial intelligence (AI) to help people express themselves and gain support from a virtual mentor they design themselves and which could be used as a left-behind avatar.

- Hidden fears: of second loss, of being found continually mourning through digital archives. For example, whilst the living person may consent to their digital afterlife existing, the loved ones of the deceased may hold a different perspective on experiencing this. Bassett (2018a) argues for the fear of 'second loss' where the bereaved fear losing the data left by the dead. Therefore, the bereaved person is faced with the original loss of the loved one and the emotional concomitants of exposure to their digital legacy, followed by the actual or potential loss of this digital legacy in the future.

Some of the more open practices include behaviours that, in the main, are unhelpful and cause distress to the bereaved. These include:

Bandwagon Mourning

Bandwagon mourners were described by Rossetto et al. (2014) as individuals who did not know the deceased very well whilst they were alive but after their death posted disproportionate commentary and pictures about them. In practice, their actions were seen as offensive because their position was not central enough to the circle of the deceased to warrant their behaviour. They were invariably seen as people who were attention-seeking and their grief was seen as inappropriate.

Emotional Rubberneckers

The concept of emotional rubbernecking (DeGroot, 2014) is defined as the actions of individuals who did not know the deceased whilst they were alive but felt a connection with them (or their survivors) following death.

In practice, they are seen as 'online voyeurs who visit the Facebook memorials of strangers or distant acquaintances to read what others write and to post their own' (p. 79). There were also lurkers who, although they did not participate, were present and watching Facebook memorial groups and the discussions on the site.

However, DeGroot found that rubbernecking behaviour was not always negative, since it was a way that some people dealt with a death that had affected them; it was used as a communal form of grieving.

Memorial Trolling

Memorial trolling (Phillips, 2015) or RIP trolling is one of the most distasteful forms of trolling; perhaps one of the most famous RIP trolls was Sean Duffy, who was imprisoned for posting messages like 'Help me mummy, It's hot in Hell' on a dead girls' page–on Mother's Day. RIP trolling is often done anonymously or through the use of a pseudonym or fake identity, but Phillips (2011) contends that most RIP trolling is not directed at the family of the deceased but rather grief tourists who have no real-life connection to the deceased. However, it is clear that there are different forms of trolling occurring in different grief spaces and clearly further research is needed in this area.

Digital Death Photography

Photos of the dead, whether pictures of corpses in funeral homes, hospitals or in coffins in church, are often posted on social media sites. Today digital death photography is public and accessible. For some people, this photography is seen as normal, for others, it is distasteful, rude and shocking. Aceti argues that:

> By constantly digitising and uploading data/lives, 'digital humanity' finds and confirms its existence as an eternal illusion and an eternal reality that constantly attempt to grasp themselves via a screen in the opposition of real vs. virtual.

(ACETI, 2015: 321)

Within the genre of digital death photography is the selfie. One of the most public and possibly surprising instances was Barack Obama, the then American president, taking a selfie with his wife and other attendees at the funeral of Nelson Mandela in South Africa.

What is interesting as Cann (2014) notes:

> The previous wearing of a black armband has been replaced with a funeral selfie in order to proclaim one's mourning status. One aspect of these selfies largely overlooked is the language and tone of the accompanying comments by the selfie posters. Many of them offer tributes to the deceased and speak of how sad they are. These tributes can help interpret the selfies. While the selfies may at first glance seem egocentric, the tributes that often accompany the pictures indicate that the photos are the final conversation between the person pictured and the deceased. These photos are a tangible way to say goodbye *in person*. That final conversation is not a mere mental memory but captured and frozen in time through photography, and perhaps it is an attempt to transcend the distance between the living and the dead.
>
> (CANN, 2014: 81)

Whilst selfies have largely become acceptable language, norms about other's posting on social media sites are often policed in challenging ways.

Grief Policing

Whilst grief policing was originally defined by Walter (2000) as a term of abuse for those who might presume to tell others how, or for whom, to grieve, a more accurate definition that reflects the digital in the 2020s would be that of Palmer (2015), who suggests:

> The grief police maintain strict yet unspecified standards of exactly who should be sad, and when, and how much. Step outside the boundaries, and they'll be there to scare you back to the safe zone.

Yet within this practice of grief policing there is the somewhat hidden practice of covert sanctioning, for example, McLaughlin and Vitak (2012) found that, within particular online communities, users were told what was allowed and disallowed and sanctions were applied for unacceptable behaviours. The difficulty with grief policing is that within different communities, the norms vary and, therefore, individuals may not know what is deemed acceptable and what is not. Gach, Fiesler, and Brubaker (2017), in light of their study, redefine grief policing as 'norm enforcement practices around grief' (p 47:5). This was based on three key findings from their study which were corrective or shaming comments that accused a participant of not grieving in a space designed for grief, policing comments that were responses to inappropriate expressions of grief and positively reinforcing comments that were regarded as ideal responses.

As discussed earlier, people post about their grief on social networking sites, but what the norms are and what is considered acceptable or unacceptable behaviour remain ambiguous. It is unclear what the long-term impact of these kinds of largely unconstructive behaviours are having on the bereaved.

LONG-TERM IMPACT ON THE BEREAVED

Many of the theories and models do not consider the long-term impact of digital afterlife on the bereaved, issues such as those set out in the following sub-sections.

The Power of Technology Providers

Kasket (2019) questions what might occur when corporations compete with us for control of these data. One public intellectual, Esther, I interviewed argued:

> The impact on the bereaved is very much mediated by the third parties, by the technology platforms that control the information and make up the rules that govern what the information is. How it is presented. Who has access to it. Who are essentially writing the terms and conditions. And who can change those at will . . . I think that because of the involvement of those third parties, whether it's a social media company or an email provider or any other kind of platform that manages and controls data, it's their involvement that means that issues around access and control and ownership and calibration of one's exposure to things are such live issues for the bereaved now . . . Like hardly a week goes by that you can't open up a newspaper and see some sort of story about it. About people struggling in their bereavement, in their grief, because of a company blocking access to something or taking something down. Culling something. Saying, sorry, you can't have access to the contents of those accounts.

It is possible that the intentions of the person predeath become distorted and re-shaped such that they no longer resemble the thoughts and wishes of that person. Furthermore, agency is called into question as the deceased is no longer able to act according to their will and agency is taken over by the corporation/s that control the data. This, in turn, raises concerns about to what ends the data might be used. It is already clear that commercial companies are using digital mourning labour. This activity is undertaken by corporate brands who use social media to share and gain from emotions of grief and nostalgia about dead celebrities to sell their products by capitalizing on grief and mass mourning (Kania-Lundholm, 2019). However, the idea of control and culling by media companies can have a profound effect, and fear of deletion is a concern for many people.

Deletion and Second Loss and Non-Deletion

Bassett (2020) compares the anticipatory fear of second loss with Victorian mourning jewellery. She suggests that just as the image fades each time the locket is opened, each time it is looked at and thought about it fades further. It is also carried around with us, just like the digital afterlives of our loved ones are carried around with us on our mobile devices. The fear of second loss creates a new form of anxiety in that it becomes a form of anticipatory grief at the prospect of losing data from a loved one. In an interview with a participant, I asked about deleting their profiles from social media or text messages, and Heather, a counsellor, explained:

Maggi: I just wondered what your view was about this kind of the keeping versus the deletion idea.
Heather: I think it's exactly the same as me deleting that other friend from my phone. I would feel it was killing them again. I mean, you can't stop people dying, but oh my goodness, you don't have to go in and kill their memory. That's weird, isn't it? But I think it would feel like killing them again. Legacy is incredibly important.

However, a different stance was proposed by Esther, who argued that it was irresponsible not to delete:

But deletion upon death is not necessarily horrible. I was really upset when Twitter cancelled its cull in November. So, in November of 2019, it announces an imminent cull of inactive accounts that was due to happen mid-December. Twitter has always had an inactive account cull policy. It's just that it's never been consistently enforced. And they have never said out in the media, we are going to do this. And there was an outcry from bereaved people. And one day later, Twitter backpedalled and said, we're not going to do the cull.

The challenge here is about the value of these tweets as well as the long-term environmental impact of storing all these data. Stokes questions whether loved ones have a moral duty to preserve someone's digital remains and questions whether deletion might count as 'harm to the deceased user' (Stokes, 2015). Further, Kasket (2019) notes that there is increasing anxiety about the preservation of our online legacy, yet as Harbinger (2020) states legal changes are occurring about digital assets but these vary considerably from country to country (this is discussed in more detail in Chapter 10).

Death Announcements by Others

Victorian mourning etiquette and the rules that governed appropriate behaviour no longer exist today, and, much of the time, it is difficult to know by their dress if people are in mourning. Further, there have been many instances when families have been usurped by social media and in extreme cases can find out about the death through the media. One stark example is that of the actor Philip Seymour Hoffman, whose death was tweeted by the *Wall Street Journal*, which broke the story that he had died in his apartment of a heroin overdose before the family had been informed. Anna, a minister in the Church of England, explained her experience:

> I have seen the negative impact of social media, particularly on bereaved families, where somebody dies and normally it would be the immediate family who announce the death in some way or share that. But it's been on Facebook before they've had chance to make the announcement themselves, because they're dealing with the deeper closer grief, but somebody else who is less connected has wanted to tell the news as it were, so that's a negative example.

Cuminskey and Hjorth (2018) explore representing, sharing and remembering loss. They suggest that mobile media result in a form of entanglement with the dead that is both private and public. What is of both interest and concern is who owns the data, what the relationship is between privacy and commemoration, and whether, since there are few guidelines, there needs to be an etiquette about how death online is managed. Shifting cultural norms towards an increase in sharing personal stories online and the expression of grief and mourning online are important to consider here.

Whilst many people are concerned about second loss and the idea of deletion, there are others who are concerned that not enough deletion actually occurs.

RECOMMENDATIONS

When considering the impact of the digital on those who have been bereaved, particularly if they are frequent users of social networking sites, it is important to:

- Be aware of regulations and social media platforms rules of conduct – if they exist.

- Locate the regulations – these are usually in the form of website privacy policies, codes of conduct and, in some cases, community guidelines.

- Encourage the bereaved, or someone close to them, to read the Terms of Service.

- Be clear about who has access to personal mourning sites – whether it is public or private.

- Appreciate that mourning norms on social media are constantly changing. Currently, inconsistent norms often appear to be the cause of conflicts and disagreements.

- Understand that although invariably the norms around death and bereavement are negotiated and established in and through social networking sites, this tends to change when parasocial grief occurs (particularly in mourning the death of celebrities) rather than in cases of personal grief.

- Realize that traditional mourning practices are often overlaid on social media sites – such as expressions of grief and remembrance being modest and understated rather than emotional and effusive.

- Be aware of the ways that implicit norms are often difficult for newcomers to a community to learn and that this can result in sanctioning that is likely to be unexpected.

The negotiation of social networking sites is complex, and whilst these are legitimate spaces to express grief, they are also troublesome spaces where those who are bereaved may require more support in their use than is perhaps initially anticipated.

GRIEF AND BEREAVEMENT IN VIRTUAL SPACES

Since the early 2000s digital bereavement communities have grown, resulting in sites, discussion forums and support groups. These include forums such as GriefNet, as well as end-of-life video conferencing and cybertherapy. However, what has also become apparent is that disenfranchised grief, which already occurred in physical spaces, appears to be increasing in virtual spaces. Disenfranchised grief refers to any grief that goes unacknowledged or invalidated by social norms or 'grief rules'. Such norms dictate who is entitled to grieve and, in turn, who receives support, acknowledgement and validation in their grief. Doka (2008: 224) defines:

> disenfranchised grief as grief that results when a person experiences a significant loss and the resultant grief is not openly acknowledged, socially validated, or publicly mourned. In short, although the individual is experiencing a grief reaction, there is no social recognition that the person has a right to grieve or a claim for social sympathy or support.

It is clear that some people use Facebook to help with disenfranchised loss. Many people using such media spaces feel validated and this, in turn, enables them to feel connected with others with similar experiences of grief. Alves (2021) notes, however, that grief is affected by culture and varies across countries, gender and time. She suggests that as we live increasingly online lives, new rituals are developing with the result that traditional funeral and ceremonial practices might be lost. She argues for the need to examine the impact that the internet and social media are having in the evolution of grief. Grief and bereavement in online spaces have grown in terms of support sites and therapy, which include the below:

GriefNet

This is an internet community for those dealing with grief, trauma and loss. It has 2 websites and 50 different email support groups and was set up in 1994 by Cendra Lynn. Lynn and Rath (2012: 96) provide details of its growth and changing practices over time. What is useful is the set of guidelines they provide for participating in the site, summarized below:

- *No flaming*

  Our primary guideline is that we be polite and respectful in responses to other subscribers. Rudeness or attacks on other people here are not acceptable.

- *Keep this private*

  Messages to this group must be kept private and confidential. Do not share messages with someone outside of this list without the author's permission.

- *Stay on topic*

  Please restrict topics to those for which the list has been created, which is your grief. Discussions of unrelated issues often confuse new members just joining.

- *No religious and spiritual discussions*

  Please do not discuss religion or spiritual beliefs.

- *Jokes*

  Use extreme discretion and post sparingly, please.

- *Limit contact with other members to group mail*

  Do not contact another member outside of the group mail. GriefNet's safety lies in the anonymity of its members.

- *No identifying information*

  Do not share your phone number, address, or anything else you would not wish anyone and everyone to know. Never hesitate to contact any of the GriefNet staff if you have concerns about someone in the group.

- *No mentioning products, practitioners and other sites*

  Products or services of any sort may not be discussed or recommended, either by supplying web addresses or by describing in detail the commercial venture.

GriefNet continues to be a successful and growing site. What is noticeable is that those who are part of the GriefNet support group themselves have become more educated about grief, through sharing the grief of others.

End-of-Life Video Conferencing

In the past, video conferencing has been used by families living far apart and this use has increased hugely during the COVID-19 pandemic. Whilst this often seems a straightforward and useful option, it can cause stress for those people less familiar with technology, and for some people, text messaging or blogging may be an easier alternative. There is a site that provides an end-of-life conferencing service: The ACTIVE (Assessing Caregivers for Team Intervention through Videophone Encounters) intervention was designed to allow patients and/or their informal caregivers to participate in meetings from their own homes using commercially available videophone technology (Oliver et al., 2009). The following other sites provide services for healthcare providers only: Project ECHO is a not-for-profit movement that aims to improve care through collaborative decision-making and problem-solving. In practice, it uses videoconferencing to conduct virtual meetings with multiple healthcare providers, undertake teaching and share good practice. 'Ivy Street' is a virtual learning environment that supports palliative and end-of-life care education, which provides support and learning experiences for healthcare workers but not for families and patients (Clabburn et al, 2020).

Whilst there is some interesting work in this area, it is clear that more developments are needed in the use of this technology.

Cybertherapy

This is the use of the internet to provide grief therapy. There are a variety of services such as BetterHelp https://www.betterhelp.com, which provide online counselling, and HealGrief https://healgrief.org, a social support network that provides a range of services including counselling. An innovative idea was developed by Landström and Mustafa (2018) who created Tuki, a mobile application (Tuki meaning 'support' in Finnish) which uses natural language processing (NLP) to analyze the situation. Tuki uses AI technology to locate a pain point in the grieving process and then matches the bereaved person to another user with similar experiences.

It is clear that in the area of death and dying, there is a place for artificial intelligence, but as yet this requires further research and development.

Certainly, Windisch et al. (2020) suggest that if palliative care centres collaborated to create larger datasets, these could provide enhanced results and imaging studies using deep neural learning which would help to provide better predictions, support symptom management, advanced care planning and improved end-of-life care.

CONCLUSION

It is clear that there is a need to consider the emotional aspect of digital afterlife creation so that we are aware of the consequences of our engagement with the bereaved on social media, not just our attitudes towards it but what we actually do. In some cases, the ongoing visibility of a deceased person's social media profile page can be upsetting to the living and seeing past conversations and photographs can result in distress and sadness. Thus, it is vital to create spaces of debate, despite cultural taboos around the discussion of death and dying, and understand the values that people place on this kind of mourning in cyberspace. Many social media sites are used in ways that ensure that mourning is not based on death and loss but instead upon the continuation of life and digital bonds in cyberspace. Yet, despite the fact that some authors suggest that religion tends to compartmentalize death and argue that religious vocabulary is not a shared construct, the next chapter suggests that this is not entirely the case. Chapter 4 considers theologies of death, the continual process of understanding and theorizing about the nature of death by those with a faith, and argues that the conflicting arguments in this area bear in-depth exploration.

Digital Theologies and Theologies of Death

INTRODUCTION

This chapter examines digital theologies and how they have been defined, adopted and implemented. The current literature in the area of digital theologies is broad with considerable overlaps between many of the different types. Further, the literature on theologies of death remains an extensive area of exploration of reflection and theorizing. This chapter attempts to bring some clarity to these ideas, as well exploring religious perspectives on media use in relation to theology. The final section of the chapter explores digital theology in relation to digital afterlife, offering perspectives from a study that examined perceptions of digital afterlife.

DIGITAL THEOLOGIES AND THEOLOGIES OF DEATH

Theologies of death are defined here as the continual process of understanding and theorizing about the nature of death by those with a faith. Theology is largely seen as the science of God, covering a breadth of perspectives about the relationship between the world, humanity and God. Translated from the Greek *theos* (God) and *logos* (word), it also comprises the study of God, the study of the supernatural and the notion of revelation.

DOI: 10.1201/9781003098256-4

TABLE 4.1   Perspectives on Digital Theologies

	Definitions	Related work
Networked theology	The ways in which religion is practised online and offline in a networked society	Campbell and Garner (2016)
Digital theology	The use of the digital to study theology	Phillips, Schiefelbein-Guerrero and Kurlberg (2019)
Digital humanities	Applying computer-based technology to the humanities	Berry and Fagerjord (2017)
Media theology	The ideological dimension of the connection between religion and media	Blondheim and Rosenburg (2017)
Digital religion	The influence religion and new media have upon one another	Campbell and Rule (2020)
Online religion	The ways in which the Internet prompted new religious practices online	Helland (2012)

Digital Theologies

Whilst a range of authors (Campbell, 2012a, 2012b; Staley, 2014; Campbell and Garner, 2016) refer to digital theologies and digital religion, there are few clear definitions of these 'theologies', and as can be seen in Table 4.1, there is also considerable overlap.

What is common to most of these definitions is the idea that the discipline of theology, understanding of religion and religious practices need to be interwoven with digital media.

A helpful overview is provided by Phillips et al. (2019), who suggest four ways in which digital theologies might be viewed. These are presented and developed further below:

- Digital technology that is used to communicate or teach theology – this is the use of media and media platforms to deliver theological education. However, such practices are already common in much tertiary education and despite Phillips' argument, this delineation seems to have relatively little to do with the actual discipline of theology per se.

- Theological research that is enabled by digitality or digital culture – as with much research today this is discipline-based research, which in this case acknowledges the use of digital media to analyse big data and evaluate areas such as online religious practice. What is perhaps different here from other disciplines is that research in the discipline of theology is generally undertaken through a close reading of inked texts; thus, the use of digital texts for this discipline is innovative.

• Engagement of theology and culture – this is perhaps a more in-depth understanding of digital theology in that it seeks to examine the relationship between theology and digital culture. In practice, the shaping of culture and theology on one another are analyzed from a reflexive position, in order to understand the impact of the digital on religious culture and practices.

• Theological–ethical engagement – this is the theological–ethical critique of digitality which seeks to evaluate the impact of technology on society through a theological lens. This perspective also recognizes the need for theologians to be operating in public spaces and exploring ethical and religious assumptions. Interestingly, much of this work is being undertaken by those in the public sphere, such as the House of Lords Select Committee on Artificial Intelligence, at which the Bishop of Oxford is in attendance, and the US in Universities that are theorizing about digital ethics, such as Campbell and Garner (2016).

It is clear that further research needs to be undertaken in this area to appreciate the impact of the digital on theology, in order to provide further understandings of the relationship between the discipline of theology and digital afterlife.

Theologies of Death and Eschatology

Theologies of death encompass issues such as ideas of belief, afterlife, liturgy, heaven, hell ritual and disposal. Davies (2008) suggests that a starting point for exploring the theology of death is Schweitzer's question of 'whether death resides inside you . . .' whether 'you have conquered it within and settled your account with it' (Schweitzer, 1907/1974: 67). Davies' theological stance towards death is that it should be seen in terms of a relationship between lifestyle and deathstyle, the idea that a theology of death must also be a theology of life. For example, if death is seen as the absurd, as empty nothingness, then life should reflect some of these attributes. In short, our lifestyle challenges our deathstyle.

It is clear that theologies of death are complex and wide ranging but can be grouped into three main areas: the problem of death, the mystery of death and the theological understanding of the mystery of death.

The problem of death – This seems both natural and absurd both for atheists and for those of faith. Death is a puzzle and as no one has experienced or really understands it, it is absurd. Camus depicts this particularly in *L'Etranger* (Camus, 1942a), which portrays the futility of a search for cohesive meaning in an incomprehensible universe without God. However, his perspective on the absurd is perhaps best captured in the *Le Mythe de Sisyphe* (Camus, 1942b), whereby Sisyphus pushes a rock up the mountain, watches it roll down and pushes it back up again in an endless cycle. Camus' point is that like Sisyphus, humans continue to question the meaning of life, but remain troubled by the answers continually falling away. However, even for those with a faith the mystery of death remains problematic, but there is also an excitation of some kind of divine revelation as a solution to this problem

The mystery of death – In several religions, the idea of the mystery of death centres on the assumption that an individuals' death is part of God's plan, and whilst God is immortal, humanity is not.

The theological understanding of the mystery of death – The wide body of literature on this topic remains controversial. What is clear is that there is little understanding of what happens to the soul after death. Some believe it is affected by prior choices made while the soul is in the state of union with the body; others suggest there may be possibilities for growth and development post-death, as discussed in Chapter 2.

Eschatology

Eschatology is the doctrine of the last things or the last times. It was originally a Western term, referring to Jewish, Christian and Muslim beliefs about the end of history, the resurrection of the dead, the last judgement and the problem of God's justice. According to the religions that have adopted the Hebrew Bible, New Testament and Qur'an, the dead either exist in a disembodied form or have a long sleep until their resurrection and the determination of their fate in the afterlife. In the Christian tradition, the end times comprise four things that are expected to be the destiny of humanity, namely resurrection, last judgement, heaven and hell. However, there is a difference in the Christian and Jewish stance towards the apocalypse: the Jewish apocalyptic stance is that, basically, God will come to a point where He will put an end to everything, and therefore the hope is to be with God, in a kind of heaven. In the Christian apocalyptic stance, there will be a new heaven and earth; heaven will come down to earth and the earth will be renewed. The central theme in the synoptic gospels is the Kingdom of God, also referred to as the kingdom of heaven, referring to power and reign rather than a place or realm. However, there are different and divergent perspectives about whether the kingdom is present, has arrived or is in the future. For example, ethical eschatology suggests that the Kingdom of God is a social entity in the present to transform and redeem society so that it has a social effect, an example of which is the Salvation Army. Futurist eschatology, the idea that Jesus prophesied about the kingdom, is in the future and has not yet come about. Realized eschatology, the idea that the day of the Lord has come, is a blend of ethical and futurist eschatology, so the belief is that the kingdom is already here and is seen in Jesus' fulfilment of prophecies, supernatural manifestation, evil being overthrown, the judgement of the world and eternal life as a present reality. Inaugurated eschatology is the idea that the kingdom is 'already and not yet'; that the dawn of the end of the age has come in the life, ministry, death and resurrection of Jesus, the Jewish Messiah, Lord and Saviour. The Kingdom of God has been ushered in, though it is not yet consummated; thus, we are living in a liminal space. The digital age might seem to be the end of an age and certainly, there is much discussion of the end times in the New Testament.

There are a wide set of traditions within Hinduism but unlike Christianity, Judaism or Islam, Hinduism has no last day or end time; thus there is not any sense or form of eschatology: no resurrection of the dead or last judgement, indeed no end of history. Instead, death is a momentary interruption in the succession of rebirths, and the law of karma determines an individual's ultimate destiny. In Buddhism, there is no parallel, there is no sense of the end times or an idea that the world will come to an end. Instead, there is *saṁsāra*, a cycle of birth and death of both the individual and of the universe, with an unending series of cycles of manifestation and non-manifestation. However, in some forms of Buddhism, there is an argument for relative eschatologies, which refers to both individual ones as personal freedom from *saṁsāra* and cosmic ones as cycles of degeneration and development, suggesting a degree of ending of phases of manifestation of the universe, or perhaps evolution. By contrast, in Islam, the Qur'an speaks of death, the end of the world and resurrection more than any other major scripture and refers to the end times which includes the last day and the day of resurrection. The 'last day' is to prepare for the day of resurrection, which could be as long as 50,000 years. Scales are used for the weighing of deeds, God himself descends to interrogate people and the scrolls of their deeds are placed in the scales, good deeds in one pan and evil deeds in the other. The ultimate consequence of hell or paradise depends on both the decision of God and the intercessions made for the dead by others.

Despite such variety across religions, it is clear that there is literature that explores theologies of death in the context of hope, perhaps implying that there is a straightforward answer to death and the afterlife, rather than the need to problematize it. Collopy (1978) provides a critical over-view of models and theologies of death, suggesting at the outset that much theological analysis of death focuses on the victorious assumption that death is the final hope and, therefore, results in the denial of death rather than an exploration of life's end. Furthermore, body and soul models of death, such as Cartesian dualism whereby the soul is released from the body on death, would seem to resolve the difficulty of death because the soul or spiritual centre of the person is deathless. However, Rahner (1975) suggests that the body-soul division is rather a thin model of death. He argues that theology must not see death as something that 'affects only the so-called body of man *(sic)*, while the so-called soul . . . [is] able to view the fate of its former partner . . . unaffected and undismayed as from above. Death affects the *whole man (sic)* the soul included' (Rahner, 1965: 30). Similarly, Kubler-Ross' stance is troublesome as she suggests that notions of truth can only really be found in acceptance (the last of her stages of accepting death), and although she argues there is darkness and struggles in the stages on the way to acceptance (Kubler-Ross, 1969), she positions the idea of a final stage of acceptance as unproblematic, which then seems to somewhat simplify the darkness of death. What is important in any kind of theology of death in the digital age is that it should not assume that death is manageable, or explicable, but also that it is contextual and needs to be fluid, as one participant, a minister and expert in life events ministry, explained:

> Do we have to look at our theology differently, particularly our the-ology of eternal life and resurrection if there is a digital . . . people can make a digital copy of themselves in the physical world? Part of theology with changing an emerging discipline and having to respond to the things that are happening in our world, we have to

constantly re-work what we think . . . I believe in life after death, I believe that my essence of me will continue into eternity, and I don't, theologically, think I need to be present in the physical world for all eternity for that to happen.

(SAVIN-BADEN, 2021A)

Although there has been relatively little discussion about the theorizing of theology and digital media, there have been studies that have examined religion and media.

RELIGION AND MEDIA

Ferré (2003) presents three views of the relationship between religion and media, which some years later has resonance, and also helps to inform some of the reasons why people from different religions respond to media and death related media in the particular ways that they do.

Media as Conduit

Ferré suggests that some religious groups see media as a conduit, a delivery system. Thus, the message is the focus and the media is deemed as neutral. This is a positive uncomplicated (if naïve) instrumental approach to media, which is seen as neutral and often as a God-given gift replaying the Christian message. Examples of this would be televangelism and the use of films, such as those by Francis Schaeffer based on his book *How Should We Then Live?* in 1977. What seems particularly pertinent from a theological perspective is that online and offline practices do actually shape one another, despite the notion of media as a conduit.

Media as a Mode of Knowing

This is an approach that sees media as value-ladened and having its own inbuilt system of knowing. This is similar to the idea of technology having its own affordances, that is, technology is not neutral but value-laden, as mentioned above. Postman has argued:

Embedded in every tool is an ideological bias, a predisposition to construct the world as one thing rather than another, to value one thing over another, to amplify one sense or skill or attitude more loudly than another... New technologies alter the structure of our interests: the things we think about. They alter the character of our symbols: the things we think with. And they alter the nature of community: the arena in which thoughts develop.

(POSTMAN, 1993: 13)

The concept of affordances has become increasingly used in research and technology since the late 1980s. The term was coined by Gibson (1979), whose idea of affordance has changed the way we understand visual perception. His argument is that the interaction of the human with the environment results in a particular course of actions. These actions are perceived in a direct, immediate way with no sensory processing, so, for example, the design of a door handle will affect whether we pull or push it. In relation to technology, the concept of 'affordances' has been used to overemphasize what particular technologies prompt or allow us to do, bringing with it a sense of covert control. It is not enough to say that a virtual world or specific type of game will make us do something in a particular way; rather we need to appreciate the way something has been designed and the way it is being used, in order to understand whether we are being made to act in a specific way by the technology. This perspective can be seen as technological determinism: the idea that those who have grown up in the digital age are necessarily different from their predecessors. This view also implies that media is all-powerful and seductive and thus people of faith need to be wary of having their values and honesty undermined. Further, this largely negative approach promotes suspicion about media use, suggesting it undermines theological values of community, justice and veracity.

Media as a Social Institution

This is where media is seen as having both embedded values and acting as a tool to be shared with users. Thus, the Internet is both a technological and a social system, so it is both a research tool and a medium of communication. This stance acknowledges the affordances of technology with its inbuilt embedded values. This view recognizes that both content and technology affect and shape users but are also context-dependent.

Social network sites seem to create communities amongst the bereaved in a way that would not occur face-to-face, for example, attending the grave to speak to the deceased. However, not only is there an assumption that the deceased live on in heaven, but there is also a belief that the dead still cares about the living and also a sense that the living and the dead will be reunited (Staley, 2014). Brubaker and Vertesi (2010) argue that expressions of this type of belief are 'techno spiritual'. Thus, social media site users tend to integrate and blend online and offline views and practices related to death.

Campbell and Golan (2011) have examined the impact of digital media and religious communities, suggesting that issues of perceived legitimacy affect the extent to which they are accepted or rejected by different religious groups. Thus, it appears that it is not the religious tradition that seems to affect technology use, rather it is the particular underlying values and goals of the given religious group. Campbell found that all religious groups she studied made deliberate choices about technology use that reflected their core values and religious views:

> Evangelicals in the United States using the internet for evangelism had much in common with Orthodox Jews in Israel using the web as a way to call secular Jews back to religious lifestyle. Both groups shared a motivation to use technology for outreach and so structured their use of the internet and work online to facilitate that primary goal. Both spoke about the Internet as an asset to their mission and as a divine resource for accomplishing these purposes.
>
> (CAMPBELL AND GARNER, 2016: 101)

Other studies, such as Morgan (2013), suggest that the performative options of new media can affect religious behaviour and rituals. Morgan suggests that what is often omitted in the relationship between religion and media is that of mediation, despite the ways in which the move from oral tradition to text has shaped the academic study of religion. Mediation is seen in religions such as Christian through the personhood of Christ who intercedes on behalf of the sinner. In Islam, there is no mediator with Allah, although people can pray for the dead as they journey and in Hinduism, mediation occurs by offering gifts to the ancestors. In the main, the focus on religious mediation has occurred through studying spaces, gender, foods, dress and forms of embodiment (Morgan, 2013). Yet, as Morgan remarks:

> New media always happen within existing ecologies, relying for their effect and appeal on the patterns they change and the attitudes and interests they challenge. The agony of learning a new

software, the congregation's unease with liturgical reform, the threat of radio, television, and cell phones to certain traditional authorities, the annoyance of teenage texting—all of these are somatic disturbances, disruptions to the inherited aesthetic of foregoing mediations.

(MORGAN, 2013, 349–350)

Blondheim and Rosenburg (2017) suggest that communication technologies complicate theology and they suggest that this is a response to the tension between the idea of God's immanence and transcendence. The challenge of their argument is the consideration of whether it is possible to use media to connect transient human beings with the unalterable deity. Blondheim and Rosenburg suggest that we now live in a postmodern world, a world with no truth which has emerged from the collapse of conventions in the digital world; this they suggest could be a new theological start. Such a new beginning could offer new religious enlightenment and create a new role for media at the interface of God and humans – but they do not explain how this might occur or what it might look like. However, Lagerkvist and Andersson (2017) argue that in a digital age, notions of gender, race and personhood are reframed through technologies. They believe that mediation should be seen as Kember and Zylinska (2011: 23) suggest, as being-in-and-becoming-with technology, essentially the idea of a continuous becoming so that mediation is life. What is perhaps more useful than the term 'mediation' in relation to digital media and religion is Barad's notion of intra-action (rather than interaction). This suggests overlap, and even integration between observer and observed, subject and object, and cause and effect, so that boundaries and relationships are created through specific intra-actions (Barad, 2003). This in turn would suggest that such overlaps should be central to different religious views and practices about digital afterlife use, as well as the relationship between artificial intelligence and digital afterlife.

ARTIFICIAL INTELLIGENCE AND DIGITAL AFTERLIFE

Artificial intelligence (AI) is usually defined as the science of making computers perform operations that require intelligence if done by humans. In the main, it seems that few people really understand the extent to which hidden AI (i.e. those collecting data which are hidden in websites) is having an impact, in terms of how information is collected and the ways in which people's lives are tracked and monitored. From the popular point of view, artificial intelligence (AI) is seen as science-fiction characters like the Hal 9000 computer from *2001*, or the androids from Channel 4's *Humans*. In modern marketing terms, it is taken to be almost any reasonably complex programme or algorithm – often based on machine-learning principles; yet, the complexity and diversity of AI are much broader than this. Recently there have been considerable improvements to AI such as better text-to-speech, improved speech recognition and high-quality avatars (the bodily manifestation of one's self). The challenge in this area though is still to cross the 'uncanny valley'; the idea that human replicas may elicit feelings of eeriness in looks, sound and especially behaviour, such as emotional responses, from something that is almost human to something that can be readily mistaken for human. There are legitimate concerns about artificial intelligence being used to control cars and weapons systems. However, it is probably unlikely, as Stephen Hawking has suggested, that 'the development of full AI would spell the end of the human race' (Cellan-Jones, 2014). In their article entitled 'A Christian Perspective on Artificial Intelligence', VanderLeest and Schuurman (2015) question what it means to be human, presenting a list of qualities, which have been paraphrased in Table 4.2.

What is interesting in their stance towards AI is that they argue that in relation to Christianity, AI raises many fundamental questions about what it means to be human which include the areas of anthropology and the study of the body, soul and mind.

TABLE 4.2   Artificial Intelligence and Being Human (Adapted from VanderLeest and Schuurman, 2015)

Human quality	Artificial intelligence and Being Human
Intelligence	Much current artificial intelligence aims to replicate both intelligence and the ability to learn, and then finally to apply that learning
Sentience	This is the ability to perceive. It is often included in research studies seeking to develop convincing virtual humans
Emotion	Emotion is not logical and is connected to both our minds and body, but it is not something easily managed in artificial intelligence
Soul	This is something that many people believe makes us human but is not something possible to test
Creativity	Being creative, inventive and innovative all seem to be essential human qualities, which are difficult to develop in artificial intelligence
Moral agency	The ability to choose would seem to be uniquely human

Harari, in *Homo Deus,* has a secular take on exploring the projects that will shape the 21st century. For those with a faith, much of what he argues is disturbing, such as the idea that human nature will be transformed in the 21st century, as intelligence becomes uncoupled from consciousness. Google and Amazon, among others, can process our behaviour to know what we want before we know it ourselves, and as Harari points out, somewhat worryingly, governments find it almost impossible to keep up with the pace of technological change. His book challenges us to consider whether indeed there is a next stage of evolution and asks fundamental questions, such as where do we go from here? Harari also seems to mix up religion and values by using the label 'religion' for any system that organizes people, such as humanism, liberalism and communism. He argues that 'it may not be wrong to call the belief in economic growth a religion, because it now purports to solve many if not most of our ethical dilemmas' (Harari, 2015: 207).

DIGITAL AFTERLIFE AND THEOLOGY

The growth of personality capture, mind uploading and computationally inspired life after death have huge implications for the future of religion, understandings of the afterlife and the influence of the dead surviving in society. Wagner notes '[b]oth religion and virtual reality can be viewed as manifestations of the desire for transcendence, the wish for some mode of imagination or being, that lies just beyond the reach of our ordinary lives' (Wagner, 2012: 4). There are a range of perspectives about digital afterlife and theology and Hutchings (2019) suggests that there are three conflicting arguments, which have been adapted and summarized below.

Digital Afterlife Is Compatible with Religion

As aforementioned, authors such as Staley (2014) and Brubaker and Vertesi (2010) indicate many people – with or without religious faith, believe that the dead can hear them through social media. Brubaker and Vertesi suggest that the dead are necessarily in heaven and they will be reunited when they die, suggesting a compatibility with the idea of life after death for the faithful:

> Many online messages are laden with the religious beliefs of the User and the friends, sometimes articulated and sometimes unsaid. For example, *I'LL SEE YOU WHEN CHRIST DECIDES*

FOR US TO MEET AGAIN . . .' and *'See you when I get to heven
[sic]'* express the specific commitment to the Christian afterlife
wherein believers will be reunited.

(BRUBAKER AND VERTESI 2010: 3)

This introduces questions about the existence and nature of the soul.
There is much debate about the idea of a soul in many religions, despite its
origins in Greek, the notion of the soul has developed through many iterations. It is now generally seen to be the mark of living things, responsible
for planning and practical thinking, as well as the location of attributes
such as courage and justice. However, perhaps what is most pertinent in
terms of digital afterlife is Plotinus' argument that the soul could not be
spatially extended, since no spatially extended item could account for the
unity of the subject of sense-perception (Lorenz, 2009). Kurzweil (1999) in
his discussion about artificial intelligence dismisses the notion of a soul,
concluding that our mind and consciousness arise entirely from the physical brain. However, Brittz (2018) asks the question about whether there
is still space in a digital age to examine the phenomenon of the soul. She
examines the visual culture that portrays technology in relation to the
soul, suggesting that in current society there is increasing interest in the
idea of the soul. The soul has its origins in ancient Greek thought, but
as she points out, African cultures connect souls and spiritualism, and
Eastern cultures see the soul as a guardian spirit. One participant I interviewed discussed the question of whether the creation of a digital immortal post-death would incorporate a soul and whether what would be left
behind would count as a soul. She reflected:

> Is it the soul of the person? I don't think it is the soul of the person.
> There's something about the physicality of a human being that is
> significant and important. I don't think it's the soul of a person.
> Whether the avatar has a soul it's like asking whether a character
> in a novel has a soul, I imagine. Does Lizzie Bennet have a soul
> because Pride and Prejudice is read by millions of people, even
> though she's not a real person, she's a creation of Jane Austen? If I
> had an avatar that lived on after my death, is it me? I don't think
> it is me. Is it my soul? I didn't think it's my soul. Is it my creation?
> Yes, it probably is my creation, but in that way, it would be like a
> character created in a novel rather than me.

Brittz concludes that through examining visual culture, the soul is prevalent in current technology and she argues that technology can manifest characteristics of the soul – but how this might be the case is unclear. What is perhaps most helpful is her stance that:

> the realm of the visual prompts us to keep considering the soul, through a variety of perspectives, such as monism, dualism and physicalism that incorporates spiritualism, religion, different cultures and a global outlook.

> (BRITTZ, 2018: 29)

Digital Afterlife Is Not Religious at All

The stance here is that although people may talk to the deceased on social media sites, they do not expect them to respond and the practice of doing this is not religious. Yet this stance seems rather stark and seems to suggest that religion is clear and bounded. Yet many people who do not attend a church or temple, as well as those who position themselves as atheists, still have a belief in an afterlife of some kind. For example, Walter et al. (2012) argue that the online dead are always accessible and often spoken about as angels. There is a sense then from Walter et al.'s perspective that angels have some kind of digital embodiment:

> Angels are messengers, traveling from heaven to earth and back, and cyberspace is an unseen medium for the transfer of messages through unseen realms, so there may well be a resonance between how some people imagine online messaging and how they imagine angels.

> (WALTER ET AL. 2012: 293)

It is difficult to see how digital afterlife (secular or faith-based) could be seen as not religious, since death discourses, certainly in the west, are imbued with religious language. Day suggests that those who believe in angels and ghosts may be atheists but that by classifying this as a religious category hides the implied value of people's beliefs about their experiences. Instead, she suggests that stories of ghostly experiences are a socially embedded 'performative ritual' (2011: 110), thus people, as it were, 'perform beliefs' in their relationship with the dead. This suggests then that perhaps the term 'religion' in the context of a digital afterlife is not a useful term to use. Staley reflects:

> Theologically, however, it seems clear that using a new media to keep memories alive and shape how the deceased are remembered is an innovation that exists alongside of, rather than replacing or radically modifying conventional Christian views of mortality and afterlife. The absence of expectation and perhaps of interest in receiving communication from the dead warns against conflating posters' belief in afterlife with posters' beliefs that the dead are still living.
>
> (STALEY, 2014: 16)

Digital Afterlife Contributes to a New Kind of Religion

The idea here is that online memorial sites become spaces of sanctuary and prayer for the living and Drescher (2012) suggests such practices may change religion itself. She suggests that this is a new 'shared theology' (2012: 215). Further, the work of Gach et al. (2017) on celebrity mourning behaviour suggests new kinds of religious practice. Campbell (2017) argues that religion can be seen in beliefs systems like Kopimism (a modern-day religion that considers file sharing and copying of information to be a sacred virtue) or Pastafarianism (a combination of pasta and Rastafarianism) which is a social movement that promotes a view of religion and opposes the teaching of intelligent design and creationism in public schools. These, and others, provide online belief systems for those with no religious affiliation. Certainly, participants I interviewed (Savin-Baden, 2021a) suggest that social media could be changing perspectives on eternity, heaven, the distance between life and death and the notion of the soul:

> in the way that telecoms and social media have killed distance, I think in a similar way it will do that with death. And in a way, as I hinted when we were talking earlier, in some ways I see a continuum between people who I've lost touch with who happen to be alive, and people I've lost touch, or haven't lost touch with, who happen to be dead. So the sort of binariness of being alive or dead somehow becomes reduced, or it just becomes part of a wider set

of contingencies about whether a person's available to me or in contact with me. So I strongly think that people will continue to exist for each other more through digital.

This participant suggested that social media has begun to replace perceived differences between life and death in a similar way to the way in which smartphones have become part of our embodied identities.

CONCLUSION

This chapter has explored the complex interweaving of theologies of death, digital theology and religious stances towards media and the afterlife. The current position seems to be one of both fluidity and confusion, with little to guide theologians, religious leaders and those with a practising faith. What is also in doubt is whether digital afterlife is more of a secular than a religious concept, even though it is evident that the use of religious language is widespread across social media sites and digital death spaces, as will be seen in the next chapter.

CHAPTER 5

Death Spaces

INTRODUCTION

Death spaces are defined here as the physical and digital spaces where the dead are seen to be located. This includes cemeteries, memorial sites, crematoria and burial grounds; it also includes digital spaces such as online cemeteries and online funerals. The chapter begins by exploring the relationship of spaces and places, recognizing in both the physical and digital realm that death spaces occur in a wide variety of forms. Analyzing the nature of places and spaces, drawing on Lefebvre's work on space and spatial practice. It then considers burial spaces, the conflict and taboos in graveyards and the development of new burial spaces. The final section of the chapter explores digital death spaces and new death spaces, suggesting that death spaces are ones that need to be interrogated continually.

PLACES AND SPACES

Following the death of a loved one, our initial thoughts are often related to the new place or space in which they might be located. At the end of the Harry Potter series, Harry meets headmaster Dumbledore, who has already died, in a space that is rather like a waiting room. Dumbledore says to Harry that he could board a train to 'go on' as he himself is waiting to go on (Rowling, 2007: 598). The notion of a waiting room and going on, or passing over to the other side, are metaphors with a strong sense of the spatial. Places too are important, whether that be the grave or other location of the remains and accident sites such as those turned into roadside

DOI: 10.1201/9781003098256-5

memorials. Kellaher and Worpole (2010) use the term 'cenotaphization' to suggest that there is a dissolving of the boundary between the living and the dead, which is discussed in more depth in Chapter 8. Such examples could include a roadside memorialization, a park bench or a Facebook page, all memorial locations where the bodily remains are seldom or never placed and thus the remains are disconnected from the site of the memorial.

Maddrell and Sidaway (2010) argue that there are particular types of spaces central to our understanding of death and mourning. These include the body as a space, in terms of the living (left behind), the dying and the dead; the domestic space of a home which may be the site of dying, mourning, private grief and remembrance; and finally public spaces such as the school, park, social club or place of worship. These spaces are physical and are often imbued with meaning and significance for the bereaved who then render these spaces as ones that are sacred. Such places and spaces remain important but so do the digital spaces. Lefebvre (1991) suggested social space might be seen as comprising a conceptual triad of spatial practice, representations of space and representational spaces.

> *Social space is a social product* . . . space thus produced also serves as a tool of thought and of action; that in addition to being a means of production it is also a means of control, and hence domination, of power; yet that as such escapes in part from those who would make use of it.
>
> (LEFEBVRE, 1991: 26)

Spatial Practice

This represents the way in which space is produced and reproduced in particular locations and social formations. This formulation of space has created spatial zones and imaginary geographies; boundaries around conceptions of time and space have moved and so we have created different kinds of 'spaces'. For example, digital mourning spaces are not seen to be as static as grave places which are bounded and uniform but instead as ongoing, variable and emergent. However, there is possibly more permeance in digital spaces than many people realize (as discussed in Chapters 2 and 3) since permanent deletion of digital photographs and social media posts are almost impossible.

Representations of Space

Representations of space are related to the relationships between sites of production and the way in which signs and codes are used within those representations. These spaces are conceived spaces and tend to be the spaces of the planners and architects, but digital architects today are companies who create and provide the platforms that are now used as sites of mourning as well. The 'real' spaces are defined by the physical world, such as the design of the cemetery, so that the values and cultures related to mourning are also sites of power and government by the state and the church. Yet, digital mourning spaces are also managed; Facebook legacy contacts have restricted rights and there are parameters on the 'Light a candle' on the Church of England website, for example. Digital mourning spaces often appear more flexible than physical ones but in fact are perhaps not.

Representational Spaces

These spaces embody symbolism, some of which may be coded, but importantly the representation is linked to what is hidden, what is clandestine. The notion of representational spaces is symbolized by activities that necessarily occur within them, whilst at the same time they embody complexity and symbolism. In practice, the representational space overlays physical space so that a kitchen may be used in the evening as a cooking and dining space but during the day as an office space. In terms of the digital, representational spaces can be seen in formulations of lived spaces and to some extent, the formulation of places created online that are designed to symbolize physical spaces. For example, during the COVID-19 pandemic, virtual marathons and virtual pilgrimages, such as along the Camino trail in Spain, became new sites of representational space. Other examples of these include death memes as well as digital slogans, murals and cenotaphization. It could be argued that Facebook memorial pages are representational spaces since they have been repurposed for symbolic use and used differently from their original intended purpose. However, until the advent of the digital, the main focus on death spaces was, and to a large extent still is, on bodies and burial.

BURIAL SPACES

Burial spaces in the UK have changed over time due to the lack of space and hygiene issues around burying the dead in the Victorian era. The result was the growth of cemeteries outside town boundaries and in more rural settings. Graveyards were not only sites of mourning and remembrance but also ones of recreation and leisure where fairs and markets occurred in Victorian times. The US had no cemeteries before 1831 but this changed with the construction of Mount Auburn Cemetery and prompted the movement to build cemeteries which comprised wide spaces and beautiful views. As in the UK, they were recreational spaces for picnics, for hunting and shooting and carriage racing. The result of these different uses, whilst now not the same in the 21st century, has in some cases resulted in conflict, concerns about taboos, behaviour and discussions about what is acceptable and what is not.

Conflict in the Graveyard

Much of the conflict in the graveyard relates to what is considered appropriate and inappropriate behaviour, although recently there has also been controversy about what is allowed on grave plots and inscriptions on headstones. For example, Margaret Keane died in July 2018, but her family were told they could not have the phrase *in ár gcroithe go deo*, meaning 'in our hearts forever' on her gravestone at St Giles Church, Exhall in Coventry, UK, without an English translation, but this was rescinded following an appeal in 2021, supported by the vicar Rev Gail Phillip. Further, there are still strict rules in the Church of England, not only about what can be written on a headstone but also about what can be left at the graveside.

Deering (2010) argues that cemeteries play an important role in both conservation and celebration but these different practices can result in them being places of conflict. In Deering's study, conflict in the graveyard was largely attributed to the behaviour of young people, with suggestions that fines should be used as deterrents against sex, drinking, graffiti and drug taking. However, Deering argues

> As young people created their own environment in the cemetery, personalising their space with litter and graffiti, they seem to be fighting back against what is described as the 'end of place' (Tuan, 1977) by locating themselves in one of the few remaining places with any sense of uniqueness and identity.
>
> (DEERING, 2010: 89)

What Deering's research suggests is that there is a need to see graveyards not as bounded spaces but as multiple-use landscapes, as was often the case in the 19th century.

Taboos and Burial

There has been considerable discussion in the last 15 years about the extent to which death is still taboo, with arguments suggesting that it no longer is, because of its continuing presence in the public domain media and particularly within and across social media. However, Woodthorpe (2010) argues against this, suggesting that this is misguided and suggesting that is an overgeneralized stance towards death, indicating that there is a disparity between academic debate and personal reality. Her research indicates that death is still clearly taboo with her findings indicating that the materiality of the buried body was not spoken about. Instead, Woodthorpe (2010) argues that death should be considered within the spatial and social context in which it is encountered, suggesting that

> the problematic status of the unbounded buried body is revealed above ground in the contestation over the ownership and integrity of plots and emphasis placed on maintaining plot boundaries. (p. 57)

One of the taboos that is rarely mentioned is the silence surrounding the dead body. Even whilst people are standing in the cemetery few people mention the buried bodies (Woodthorpe, 2010). Unlike the clear development of social media spaces for the dead and a belief in the dead residing in cyberspace, people visiting graves maintain silence about the buried body – there was a feeling that they were 'there and not' there (Meyer and Woodthorpe, 2008), in the sense of being liminal. Thus the use of euphemistic language coupled with the lack of acknowledgement of decomposition and, at the same time, the need to maintain the individuality of a grave provide clear indications of death still being taboo in the 21st century.

Bodies and Burial

In the 21st century, most people in the UK are cremated. A YouGov survey in 2016 (Smith, 2016) found that 58% of people want to be cremated compared with 17% of people wanting to be buried. However, there continues to be a struggle to bury people in graveyards and cemeteries because of the lack of space. This is further compounded by the law, which prevents the multiple uses of burial sites so that multiple bodies are not allowed to be placed one on top of another in a grave. A recent development has been the introduction of upright burials, a collaboration of the DeathTech Research Team at the University of Melbourne and the Managing Director of Upright Burials (Gould et al, 2021). Not only does the project examine the implementation of upright burial but also it examines the vertical stacking of remains above the earth in high-rise structures developed due to a shortage of land allocated for cemeteries.

The demand for burials has also caused is an increase in the creation of burial sites on private land. This is one of the death spaces that is rarely researched or discussed although it is not actually a recent practice. Research by Gittings and Walter explored the spatial and temporal complexities associated with burial on private land; they argue:

> Spatially, there are three possible locations for a private grave: 1) the garden immediately near the house, 2) a more remote part of the garden, and 3) elsewhere on one's own estate or on someone else's land. Where resources allow, the second and third more liminal locations were and are preferred, for practical, emotional and symbolic reasons and historically these graves have proved less likely to be disturbed.
>
> (GITTINGS AND WALTER, 2010: 95)

What was evident in their historical study was that those choosing to be buried away from consecrated graveyard spaces had often not anticipated that their dead body might be moved. However, in the 21st century, this seems to be less of the case with the advent of natural burial grounds. Clayden et al. (2010) explored the growth of natural burial grounds in the UK. The study found that the attraction of using natural and woodland burial sites related to the social and physical landscape of the spaces. Whilst most natural UK burial grounds in the 21st century are provided by local authorities, there are an increasing number of independent providers which includes farmers, landowners and charitable Trusts with the most prevalent being farmers. Farmers have opened burial grounds on their land both as a means of gaining income and for emotional reasons so as to allow other families to hold a more personal funeral than perhaps the farmers had experienced themselves for their own family at the traditional burial sites. What was evident was that the social spaces as well as the physical space were seen as important to farmers and families using the burial site. These social spaces were about developing a relationship with the farmers as well as with the land illustrating that these kinds of death spaces are becoming social spaces. This is exemplified through the farmers providing support for the bereaved, such as planting bulbs; the farmers and their families seeing it as an emergent landscape and the link that these sites provided with a personal history of those providing the burial space. Further, some of these new burial grounds also allow the use of technologies such as QR codes and holograms that traditional graveyards do not.

DIGITAL DEATH SPACES

Digital death spaces are defined here as those digital spaces the dying create for the living or the living create to remember the dead, in which communication and interaction are assisted, created or enhanced by digital media. Digital spaces demand that we confront the possibility of new types of visuality, literacy, pedagogy, representations of knowledge, communication and embodiment. Thus, as Pelletier has argued, 'technologies are systems of cultural transmission, creating new contexts within which existing social interests express themselves' (Pelletier, 2005: 12). Brubaker et al. (2013) suggest that social networking sites could be said to expand the nature of public mourning through temporal, spatial and social expansion, as summarized below.

Temporal expansion refers to the increase in and *immediacy* of death. In practice, the users discover the death of friends and can post post-mortem tributes soon after death, often within hours of death. The expansion this provides means that memories and updates can occur over time as they are edited and curated. The result is a temporal shift from the final rite of passage of the funeral to the ongoing grieving into everyday life.

Spatial expansion describes the dissolution of geographical limitations. This enables people from all over the world to interact, which, in turn, results in the spatial expansion of the social processes around death and bereavement. It also creates participatory mourning as well as opportunities to watch the funeral service online from afar.

Social expansion is used to describe the way in which social networking sites are used to share information across diverse groups of people. This includes friends and relatives, who may not necessarily know one another but are united by their love of the deceased, and it is the deceased who become the centre of the social expansion. Thus, digital spaces are not just virtual spaces but are spaces affected and created by diverse media. However, the ways in which digital spaces are used pre- and post-death are complex and in some cases troublesome in terms of overt and covert practices. Digital death spaces may replicate physical spaces, but they also have their own language and semiotics.

There is a sense now that death in the media and death in digital spaces at times is taking on a larger reality than everyday life – the virtual has become the real so that 'rather than art imitating life, life imitates art. One's social currency begins to feel *real* and the importance of presence becomes neglected' (Cann, 2014: 76). Thus, replication occurs in digital spaces through symbols such as crosses, angels and memes, as well as poems and maxims. The physical images of death are transposed onto digital platforms. The result is that the reproduction of memories, photographs and occasions on social media does appear as if the image 'being repeated no longer refers to things themselves but to their reproductions' (Foucault, 1987: 22). It is as if death and death memories are being reproduced and replicated all the time, constantly all over the world in an endless cycle of replication. This re-fabrication results in diverse perspectives, copies, performances, enactments and artefacts. Whilst there are often pictures of dead loved ones posted on Facebook and Instagram, rarely are there pictures of graves and graveyards. Therefore people using social media and the social medial platforms themselves do appear to be changing the language of death and loss.

Mori et al. (2012) examined the affordances of three online platforms: MySpace, YouTube and an online condolence book. The study examined these spaces after the death of the murdered American 17-year-old Anna Svidersky by a stranger in a fast-food chain in 2006. The affordances they examined were

- Persistence: the continuation of comments five years after her death
- Replicability: the repurposing of digital content after death
- Scalability: the sheer number of audiences participating online

- Searchability: the ability to search or explore the past because net-worked technologies support tagged information

What is particularly useful from this study, even a decade later, is that the authors suggest the need for internet companies to consider language and semiotics in the ways that sites are designed. Mori et al. (2012) argue that death needs to be treated differently online. They suggest the need for companies to create new tools to enable the repurposing of content from social networks to support memorialization, as well as the provision of a moderation facility by service providers to reduce trolling of the bereaved. It is clear then that people's different expectations about the purpose and value of a space, themselves within that space and the semiotics and language used all affect the ways in which they participate in/on a given platform.

MOURNING SPACES: GENRES OF PARTICIPATION

Mori et al. (2012) argue that social media platforms affect the style and content of what people share. For example, Instagram, unlike Facebook, does not have a space where mourners can share grief; therefore, they are more likely to post on a sharing site where collective mourning can occur. These genres of participation are similar to the work of Ito et al. (2010), who argue for genres of participation seen in young people's relationship with media, which they define as a 'constellation of characteristics' that are constantly changing. Ito et al.'s terms of hanging out, messing around and geeking out have been adapted here to reflect the mourning genres of participation.

Hanging Out

In the study by Ito et al. (2010), hanging out involved fluid movement between online and offline activities. In grief spaces, mourners are often seen hanging out and posting on Facebook. In 2020, Moyer and Enck investigated the reason that people post grief messages on Facebook and what they felt about others' grief posts. The findings indicated that the primary reasons were for commemoration, connection, expression of emotions and remembrance of occasions.

Exploration

In the study by Ito et al., exploration was characterized by teenagers being involved in intense engagement with new media, rather than being particularly friendship-driven, involving both 'looking around' as well as 'experimentation and play'. In grief spaces, experimentation occurs and a poignant form of exploration is the selfie where individual grief and the memorialization of the dead occur at the same time. Selfies are seen with mixed views because they bring both death and mourning into homes yet are also are seen by some people as being disrespectful. For those who have lost a loved one a selfie is invariably used to garner support since they post selfies of themselves dressed for the funeral displaying sadness. The ensuing likes and supportive posts then confirm them in their grief.

Geeking Out and the Use of Expertise

Geeking out is defined by Ito et al. (2010), as intense engagement with technology or media and rewriting the rules, which also occurs in grief spaces. Moyer and Enck (2020) suggest that digital postings about grief serve to create and maintain a relationship with both the deceased and other mourners. The intense engagement is characterized by communicating directly with the deceased's social media sites and pages, using hashtags to ensure messages can be aggregated, and by tagging the deceased in posts, so that there is an ongoing connection with the deceased's profile pages.

NEW DEATH SPACES

New death spaces are often envisaged as emerging just from the digital, but other new death spaces have developed. One such space is the Death Café, which offers people the chance to meet, drink tea, eat cake and discuss death. The origins of Death Café began when Bernard Crettaz and his wife Yvonne Preiswerk in Geneva, in 1982, were asked to examine funeral rights and customs related to death in society. After her death, Crettaz organized *Café mortel events* to talk about death at the bistro. Death Café (https://deathcafe.com/) was based on this work and set up in the UK in 2011 in East London by Jon Underwood. It was set up to not only increase people's awareness of death but also help them to make the most of their lives. It is termed a 'social franchise' so that those who set one up and use the guide and principles can use the name Death Café and post events to the website. It is a not-for-profit organization that offers discussions about death, rather than counselling.

In terms of digital death spaces, these tend to differ not only according to the platform but also the ways in which they are used, varying according to the users' ages, cultures and belief systems. The result of such breadth is that they seem to have a liminal quality, they are betwixt and between spaces. This is partly because such spaces are digital, but it is also because they are death spaces, spaces which are about the dead and also where the dead are assumed to live on. Such death spaces are also liminal, which is discussed in more detail in Chapter 7, because they are constantly changing; the rules that govern them are constantly on the move and the practices within them are unclear and inchoate.

The areas of death spaces and digital memorialization tend to overlap with one another. For example, people join and engage with the World Wide Cemetery, Virtual Memorial Garden or Legacy.com. The worldwide cemetery (https://cemetery.org/) was created by Mike Kibbee in 1995, and it is the oldest online cemetery and memorial site in the world. In practice, a one-off fee enables the creation of a customized, permanent memorial in any of six languages. Whilst the site says you will have a permanent web address, it is not clear what will happen to this site over time, or indeed the memorials on it. However, the site states that there is sufficient money in a dedicated bank account to cover hosting costs for 100 years. The Virtual Memorial Garden (http://www.virtualmemorialgarden.net/) is a free site, which is much more basic, where there is a guest book and the possibility of leaving virtual flowers. Legacy.com (https://www.legacy.com/) is a comprehensive site that celebrates life and has a focus on celebrity death on its home page. It states that it is 'the place where the world pauses to embrace the power of a life well-lived . . . is the global leader in online obituaries, a top-50 website in the United States, and a destination for over 40 million unique visitors each month around the world'. It provides is a series of helpful articles and planning resources.

A more recent development is the Encyclopaedia of Cemetery Technology https://cemeterytech.omeka.net/ – a global directory of technologies that enhances the experience of visiting or interacting with a cemetery (often aptly referred to as Cem Tech). It was created by the DeathTech Team at the University of Melbourne as part of the Future Cemetery project. This site provides a wide range of information and resources as well as links to other interesting digital artefacts and apps such as the Cemetales app, a mobile app designed to facilitate a self-guided audio tour of the Common Cemetery in Aalborg, a crowdsourced online database of graves in Canada, a sound system designed to play music inside a buried coffin and a design for a gravesite holographic projection system.

The use of digital media related to funerals and funeral rituals remains conservative and inconsistent. Currently, digital media practices are largely mediated by funeral directors who are often family businesses with traditional values about how death should be managed. However, since the global pandemic, the number of online funerals has increased beyond any prediction and this has brought with it sites such as Memorable Words (https://www.memorablewords.co.uk/online-funeral-services), which can design the funeral for you online, create a live zoom link to watch it and also provide a post-funeral video. Other companies such as funeralOne (https://www.funeralone.com/) offer website design, funeral webcasting, a memorial website and funeral tribute video software. There are also what are termed 'drive-through funerals' (which are in effect body viewings) where there is a screen which enables people to drive past and view the body at the funeral home. Another company is Hugs From Home, which provides a site allowing friends and family to send a message of condolence for the grieving family, which is then handwritten and attached to a white balloon which is placed in the crematorium. Whilst as yet there is little research into the impact of online funerals, what is becoming clear is that they are creating different forms of social but private mourning spaces.

Funerals held during COVID-19 in the UK resulted in severe restrictions on the number of onsite mourners, initially for the first 2–3 months limited to less than 10 people and later to 30. Therefore, many mourners watched lived-streamed or online funerals in the privacy of their own home, so that the funeral became both a public and a private space. Accounts indicate that many mourners watching the funeral online felt as if it was voyeuristic and they were not really part of it. While online funerals taking place for those who knew the deceased have caused and added a sense of loss with the feeling of not being able to say a physical goodbye, the situation with the death of celebrities has been different. Here, the development of parasocial relationships in online spaces has resulted in fans believing they are experiencing legitimate grief, and online funerals tend to strengthen the sense of legitimate grief.

INTERROGATING DEATH SPACES

It is clear from the studies in the area of death spaces that there continue to be mixed views about the extent to which death persists in being a taboo and whether it is actually sequestered in the 21st century. Much of the debate suggests that death has re-entered the public discourse, largely through social media. However, whether the visibility of death and the dead has resulted in less taboo and sequestration remains questionable. What is evident is that views about bereavement and the creation of media-based bereavement spaces have changed. This is reflected in the increased usage and number of death-related social media sites, in the shift in burials to more open and environmentally friendly sites and in the open discussions about dealing with death in online spaces.

Alternative forms of death spaces can be seen not only through dark tourism as will be discussed in Chapter 11 but also through the presentation of death in exhibitions and art. For example, 'Death and the Human Experience' was an exhibition in 2016 that sought to enable people to start a conversation about death. Held at Bristol Museum and Art Gallery in the UK it included diverse objects from Victorian mourning dress to a Ghanaian fantasy coffin, as well as symbols of death, stages of death and human remains. It also provided fascinating stories from cultures across the world. It was designed to encourage visitors to consider ethical issues, different attitudes to dying and how different cultures deal with death and is now online (http://museums.bristol.gov.uk/narratives.php?irn=1 3459). However, the exhibition of preserved human bodies, 'Bodyworlds', provoked controversy (Mao, 2018). Created through a process known as plastination, 20 cadavers were displayed and argued to be art, science and emotion as a museum of the self.

It is clear that there is an emerging interest in exploring death through new and diverse spaces, whether gentle discussions in cafes or arresting exhibitions. It seems to be the case that death spaces are heterotopias, spaces that are not only other-worldly but worlds within worlds that are not only mirroring but also upsetting what is outside. Johnson suggests:

> Cemeteries can be said to incorporate many of the features and principles that Foucault lists as characteristics of heterotopias; for example, they recall the myths of paradise; they manifest an idealised plan; they mark a final rite of passage; they form a microcosm; they enclose a rupture; they contain multiple meanings; and they are both utterly mundane and extraordinary.
>
> (JOHNSON, 2013: 799)

Heterotopias emerge in different facets and phases of our lives; they are sites and spaces that invert, distort and unsettle. Foucault's delineation of heterotopias ranges from prisons and asylums to brothels and baths. These sites are seen as unsettling spaces because they illuminate the extraordinary in the ordinary. This is evident in digital death spaces as well as in cemeteries, where memorialization is central and the body is ignored. Yet, at the same time, these death spaces are not sequestered but coexist with everyday life and perhaps might be related to the idea of a fold. Folding means there is disruption between the idea of an inside and an outside so that inside and outside are both inside and outside; to reiterate, 'a fold is always fold within a fold' (Deleuze, 1993: 6). Death spaces, of whatever sort, interrupt life spaces whilst also providing, at the same time, a refuge in a suspended 'not normal' space.

CONCLUSION

This chapter illustrates that in both physical and digital death there has been an expansion and change in burial, bereavement practices and the use of social media. Death spaces are ones that provoke a need to explore binaries; they are spaces that are open and hidden, above and below, and private and public. This sense of almost overlapping binary results in an increasing sense of the liminal will be explored in Chapter 7. Whilst death spaces are unsettling and uncomfortable, they also appear to be spaces of fascination. There is a sense of intrigue about whether the dead might be located, and if indeed they can be located, resulting in digital death spaces and physical death spaces being places of contradiction both temporally and spatially, hence prompting questions about where exactly the dead are located in space and time. Yet, what also remains an area of confusion for many people is the idea of what might count as a good death and how this might be managed, which will be explored next in Chapter 6.

A Good Digital Death

INTRODUCTION

This chapter begins by examining the notion of a bad and good death in order to contextualize the idea of a good digital death. It is argued that planning, preservation, mediation, transference and taboo management are central concepts in the administration of a good digital death and ways of achieving this are suggested. The second section of the chapter presents areas that can affect not only the possibility for a good death but also good mourning in terms of taboos, etiquette and unintended consequences. Thus a good digital death relies both on the practical tasks of creating a digital legacy and on the effective management of the evolving dead. The chapter begins, however, with an exploration of the notion of bad death.

BAD DEATH

In order to contextualize the idea of good digital death, it is first important to explore the whole idea of what might constitute a bad death. To date, there has been relatively little research into the idea of a bad death. Some of this work was undertaken by Jacobsen (2017) who explored the notion of a bad death, suggesting that this is an area that is largely overlooked. The bad deaths, he describes, comprise dying unexpectedly, dying unprepared/unresolved, dying painfully, dying alone and dying undignified. Jacobsen's five types are summarized.

DOI: 10.1201/9781003098256-6

Dying unexpectedly. When someone dies unexpectedly it is a shock to family and friends and it makes it impossible to say farewell. In many cases, there is no sense of resolution or completion, with the result that there are many practicalities to deal with which are difficult to manage. This is the kind of death that seems to deprive the dead person and their relatives of the chance to make that death memorable or meaningful in any way.

Dying unprepared/unresolved. This overlaps to some extent with dying unexpectedly. Having an unprepared and unresolved death is often about not having said goodbye and not being emotionally ready to die and to leave the others behind. The process of being prepared for death and having issues resolved prompts people to see the death of their loved ones as part of their life. Thus unprepared/unresolved death prevents or interrupts relatives and the dying from facing death as part of life. Yet many people today say they would prefer a sudden death because of the lack of suffering it brings; however, at the same time, medical advances which are seen as welcome have also become the cause of a bad death because people linger on.

Dying painfully. The management of pain is something that has been discussed widely in the palliative care movement, so part of a good death is often seen as the good management of pain. For most people, the management of a painful death is something that they want to avoid; a bad death then is a painful death because pain diminishes people and makes them feel vulnerable and not in control of the dying process.

Dying alone. This is a bad death if it is perceived to be so by the person who is dying. Many people may choose to die alone and even wait for their visiting loved ones to leave the hospital or hospice before they die. Yet for some people dying alone is seen to be a sign of failure since dying alone means that the final journey is undergone without the presence of loved ones and results in it being a lonely affair.

Dying undignified. Dignity is something that remains a hotly debated concept. In terms of death, this is bound up with how people die, where they die, the way they are treated and whether they are alone or not when they are dying. Thus undignified dying is classed as a bad death since it does not respect the humanity of the dying person. The undignified death is perhaps one of the most feared by people in western society because it means leaving life without autonomy, without support and without the dignity that is seen by many as so central to a good death.

Jacobsen points out that death is not only a physical reality and even a technical phenomenon but also a social construction, because the understanding of death has changed. The result is that today we do death differently from former years and thus as Jacobson says, death is:

> something contextual and something that is historically, culturally and socially constructed by people as an integral and continuous part of their being-in-the-world.
>
> (Jacobsen, 2017: 353)

This work by Jacobsen sheds light on the issues relating to a bad physical death, but as yet there seems to be little, if any, research on the idea of a bad digital death. Most of this small body of research seems to relate to the extent to which people have managed their digital legacy well and in 2021 this is something which still appears to be in the shadows of people's lives. What is apparent is that digital bad death seems to be related to four main concerns:

- Unexpected loss of digital legacy – This can occur in a number of ways but the most common is when someone dies an unexpected death and the consequence is that there has been no preparation for the digital assets in the will and so the digital legacy is lost. This is what also occurs when people make an informed choice not to include their digital assets in their will or opt to nominate a legacy contact on social networking sites.

- Accidental loss of digital legacy – This occurs when data is held on a device, typically a mobile phone and due to changes and updates, data are lost. An example of this is provided by Hård af Segerstad et al. (2020), who narrate the story of a widow who charged and paid for the mobile phone of her late husband. The family used WhatsApp and he was still included in their group, but the failure to update the app resulted in the sudden departure of the deceased from the WhatsApp group to the consternation of the family: 'As a result of design and function (built in obsolescence), the technological apparatus that sustained his digital existence was superseded with new iterations resulting in his digital demise' (Hård af Segerstad et al, 2020: n.p.).

- Unprepared digital legacy – This is where someone wanted to leave their digital assets clearly marked for their loved ones to retain, but this has only been undertaken in a nominal way. The result is that those left behind may not have access to digital photographs, music and social media, or only parts of it.

- No digital legacy provision – This is when no provision has been made by the deceased, either by choice or by accident. For those left behind this can result in loss of artefacts they wish to retain, due to them not possessing any legal right or ability to obtain the artefact.

Ways of managing or avoiding a bad digital death will be explored later in the chapter. However, in order to understand what might constitute a good digital death the idea of a good death in itself will be explored first.

A GOOD DEATH

The idea of a good death emerged from the Middle Ages when a good death was seen as one that guaranteed passage to an afterlife through different kinds of rituals (Aries, 1974). Today the notion of a good death is more about saying goodbye and having our affairs in order. A good death is also linked to being in control of our dying narrative. This may include the idea of a bucket list but also planning a correct balance of factors such as choice, autonomy and time with loved ones. Yet there still remains controversy and diversity of opinion about what constitutes a good death. For example, Meier et al. (2016) undertook a literature review and identified 11 themes of good death with the top 3 being preferences for the dying process (94%), pain-free status (81%) and emotional well-being (64%). However, there were discrepancies between respondents' views and family wishes, with the family focusing more on quality of life, completion and family involvement. Drawing on the work of Sandman (2005), I would argue that a good death is one which explores and deals with the following three chronological situations: the period of dying, the event of the death and the post-death situation. A good death would then mean it was about making the person whose death is imminent comfortable, and as least devastating to their family as possible; then helping the grieving family to come to terms with the death; and finally the proper management of all the post-death arrangements and the subsequent bereavement. There are many sites that offer guidance on this today and they include issues such as whether people want to be told when they are close to death and who should be told they are near the end, what the final day might look like and who should visit during this time. What is apparent in the management of physical death is the autonomy of the one dying, respect and a sense of a useful legacy. It would seem these issues are also important in the administration of digital death too.

CHARACTERISTICS OF A GOOD DIGITAL DEATH

Whilst there is a considerable body of literature on the idea of a good death, there is very little on the idea of a good digital death. A good digital death does perhaps include similar attributes to that of a good physical death, such as autonomy, choice and discussion with loved ones (see Table 6.1). Most current research merely focuses on grief and mourning, with little research exploring the socio-cultural and socio-political impacts. Further, there is little evidence about the impact of eternal endurance and instant vanishment on recipients, family, friends and religious leaders, nor its

TABLE 6.1   Characteristics of a Good Digital Death

	Definition	Types of activities
Planning	Planning the digital death and burial of the deceased	Virtual will for the digital self, creation of a death design protocol
Preservation	Those wishing to preserve the memory or the form of the deceased digitally	Memory creation, virtual human creation, virtual persona creation and posthumous career planning
Mediation	The use of professional support to ensure that the digital legacy is created and all those involved are in agreement with it	Working with families and friends to ensure agreements, discussing digital death etiquette and unintended consequences
Transference	The agreement with loved ones about what will be left behind by whom and for whom	The creation and leaving behind of a digital will, digital artefacts and a digital legacy
Taboos	The management of the taboos surrounding transference and preservation of the digital legacy	This comprises a number of activities such as covert actions that may be seen as unacceptable to a wider public, such as using the deceased's phone, posting messages on social media as if it were the deceased speaking or posting inappropriate comments on memorial pages

effect on mourning practices. Pitsillides et al. (2009) suggested that digital death can be seen as either the death of a living being and the way it affects the digital world or the death of a digital object and the way it affects a living being.

Planning

Pitsillides et al. (2009) propose the following three ideas which I suggest could be used for planning a good digital death:

1. Virtual will for the digital self.

   This could occur through Facebook, but also through other services that offer guidance on digital will creation. Pitsillides et al. also suggest the idea of 'D-mail' or death news mail that could be a service that provides a pre-agreed wording announcing the news of one's death which would be posted to both virtual and physical friends. As yet, this does not appear to be a service provided by any companies, but it would seem to be a helpful idea.

2. Death of Information.

   A service of visual communication to comfort the 'grieving' owner whose personal computer has died could be considered. This idea is about the loss of a digital legacy, the 'things' on the computer. Pitsillides et al. suggest the idea of the 'burial' of the dead computer.

3. Immortality of Information and the need to create death.

   Even though this work was written over 10 years ago (Pitsillides et al. 2009), the suggestions here remain pertinent and innovative:

   • Design an agreed protocol, or service, which will 'kill' all information that is no longer 'alive' such as search engines that tag information that is dead.

   • Design a digital archaeologist – the idea here is that similar to physical life, in the digital realm there is an archaeologist who harvests data, archives it and decides what is waste and rubbish.

   • Design an online funeral service – the idea of the online funeral service has now become commonplace, although few people actually design these before they are dead. An online funeral service as suggested by Pitsillides et al. would identify all the digital

possessions of a deceased person and cause the deletion or modi-
fication or archiving or attributing of this information, according to the instructions of the family or the will of the deceased.

- Design a service to 'recreate' or clean a computer – this service
largely happens through reimaging computers today, although,
again, it was a good suggestion in 2009.

- Design a service which would aid a computer in 'committing suicide' after the death of its human.

Few people today plan their digital legacy or digital death, but as the
growth of the data of the dead continues, there needs to be a solution and
one that could perhaps be developed through artificial intelligence or
machine learning.

Preservation

Preserving oneself or being preserved by someone else may affect both the
dying person's peace of mind and the well-being of the bereaved. Further,
it is not clear whether the possibility of a digital afterlife and the use of digital media alter thoughts about the mind–body connection and whether
interaction with a person's digital afterlife alters one's spiritual journey
through grief. Those that are part of the preservation process can be either
those considering their own digital preservation or those who preserve
someone digital after death. The latter may use memories and artefacts to
create a digital legacy, such as a memorial site or digitally immortal persona and may include the deceased themselves before death. Three potential types of preservers can be identified.

Memory creators: Those creating passive digital memories and artefacts pre and post the subject's death. These have already been considered
above in examples such as virtual venerations, digital commemoration,
digital memorialization and enduring biographies, as mentioned in
Chapter 3. Memory creators are invariably those who are memorializing
the deceased. However, there is an increasing trend by those facing death
to begin memory creation and memorialization while they are still living.

Virtual human creators: Those creating a representative interactive virtual human predeath. This is usually created by the person in conjunc-
tion with friends and relatives. The current status of post-death virtual
humans (as will be discussed in Chapter 8) remains quite limited: they are
able to conduct a limited conversation with others but they have a limited

capability to learn, grow, act on and influence the wider world around them (and hence could be considered a virtual humanoid in the typology identified by Burden and Savin-Baden, 2019). To date, such a virtual human would have a minimal likelihood of being mistaken for a still-living subject. An example of this is the virtual persona created by Daden and discussed in more detail by Burden (2020).

Persona creators. This refers to those people who seek to create a digitally immortal persona predeath that learns and adapts over time and can influence and act on the wider world around it. Hence could be considered a virtual human or even ultimately a *virtual sapien* (Burden and Savin-Baden, 2019). It has a high likelihood of being mistaken for a still-living subject, but as yet these are still in quite the early stages of development.

Posthumous career planning. This is a new form of preservation that is occurring which is being led by the film industry. A firm entitled Digital Domains works with actors to enable their digital legacy to continue post-death. It is possible to archive not only film footage that can be inserted but also voice, body movements and expressions. Penfold-Mounce and Smith note that there are huge financial possibilities for earning post-death. The Death Rich List illustrates the impact of this industry as the authors suggest:

> Writing and recording songs in life enables the person to keep performing and earning after death. Death is far from the end of a career; it is just a diversion towards a different manner of working.
>
> (PENFOLD-MOUNCE AND SMITH, 2021: 44)

It is important to note, however, that all these preservation activities and processes do require mediation and management by those left behind.

Mediation

Mediators are professionals, such as bereavement counsellors, religious leaders or lawyers, who support both people predeath to create digital legacies of themselves and the bereaved who receive or encounter legacies, memorial sites or digitally immortal persona. An understanding of multiple perspectives, including the wishes and desires of the person predeath and the potential idiosyncratic response of the receivers to experiencing digital afterlife of the deceased person, is becoming more important to this group. Important issues to consider would also be digital death etiquette and unintended consequences. Whilst we are a long way from Victorian mourning etiquette and the rules that governed appropriate behaviour, there do seem to be some behaviours that need to be mediated. Wagner (2018) reviewed a range of studies exploring mourning online and has suggested that norms are constantly changing and being renegotiated by users of social media. A distinction is also made between those who are mourning themselves and reactions to the mourning of others, highlighting the requirement to consider the needs of both mourners and the users of social media who are reacting to expressions of grief online. Cuminskey and Hjorth (2018) suggest that mobile media result in a form of entanglement with the dead that is both private and public. What is of both interest and concern is who owns the data, what the relationship is between privacy and commemoration, and whether, since there are so few guidelines, there needs to be an etiquette about how death online is managed. Shifting cultural norms towards an increase in sharing personal stories online and the expression of grief and mourning online are important to consider here.

Kasket (2019) introduces questions about the unintended consequences of the digital on death, suggesting that the dead continue to 'speak' themselves. Further, she asks pertinent questions about how we manage our own data and what might occur when corporations compete with us for control of these data? It is possible that the intentions of the person pre-death become distorted and re-shaped such that they no longer resemble the thoughts and wishes of that person. Hence, the agency is called into question as the deceased is no longer able to act according to their will and agency is taken over by the corporation/s that control the data. This in turn raises concerns about to what ends the data might be used. It is already clear that commercial companies are using digital mourning labour. This is an activity undertaken by corporate brands who use social media to share (and gain from) emotions of grief and nostalgia about dead celebrities in order to sell their products by capitalizing on the grief and mass mourning following the death of celebrities.

Transference

Transference is the process of leaving behind the digital legacy. Whilst the notion of transference here is used to mean the transferring of the digital artefacts from the deceased, it also recognizes that from a psychological perspective transference involves the redirection of feelings about a specific person or event onto someone else. Thus whilst loved ones and friends will receive the memories and artefacts, including memorial sites, digital messages and/or digitally immortal persona, many of these will be imbued with underlying feelings and emotions that are perhaps generally not recognized in the digital transfer of a digital legacy. Thus, the reactions to the creation of digital memorials or to a digital persona may be highly idiosyncratic and unpredictable and indeed may change over time.

Taboos

Freud (1918/1963) argued that we have attempted to eliminate death from life. Those who have been studying death and working in palliative care have argued for many years that death is the last taboo in modern society. Despite the media interest in death and the growth of the dead on social networking sites, there are still taboos associated with death, despite claims that this is not the case. Further, there are suggestions because of the existence of social networking sites that death is no longer sequestered (Walter et al, 2012; Kasket, 2021). I would argue that in many ways it still is. For example, the spectacular deaths occurring in films make death seems to be something that largely happens to young heroes and heroines, and the medicalization of death and medicide mean that death is sequestered but in different ways than in former years. Clarke notes that media portrayal of death in magazines suggest that death is

> often a matter of *individual freedom*, the *result of personal preference* and thus potentially or actually *under control*. People can elect to suicide to extend their lives, to use euthanasia as they will. Death is not random, unwelcome or to be feared. It is not portrayed in the context of prevention, suffering, palliation or community supports. The links between economic, ethnic and other forms of inequality and death rates are ignored.
>
> (CLARKE, 2006: 162, italics in original)

What is particularly interesting about Clarke's study is the indication that the media focus for managing death is through medicine rather than artificial intelligence technology. Thus, the idea of 'medicide', that is death through medical intervention, suggests that the societal view is that death needs to be chosen and managed by the individual whilst still alive. What is also of interest is that Clarke notes the absence in media portrayal of the death of:

- discussion of poverty violence, racism inequality and international conflict

- death occurring naturally amidst family and friends

- death described as a positive end of life journey

- the role of religion in death

These findings illustrate that media portrayal of death obfuscates the complexities and pain of death and loss, resulting in a sense that individuals have the right to control their own death on their own terms. This perhaps also reflects the desire to use artificial intelligence to manage death through the creation of post-death avatars.

Yet there is, at the same time, a redomestication of death resulting in new traditions and personal rituals (Jacobsen, 2021: 5). Examples include the management or enshrining of ashes, public roadside memorialization and the creation of digital photographs where the living are inserted into pictures of the dead. However, many taboos are covert, hidden in the behaviours of what is acceptable and unacceptable in the management of death in social media spaces. For example, Sofka outlines netiquette relating to digital legacies and dealing with death, tragedy and grief (Sofka, 2020b). Within Sofka's work, there are a series of suggestions which I would argue are taboos; some of these are taken from her extensive list and summarized below:

- Do not request changes to the deceased's social media platforms or online services without the express permission from either the deceased before they die or their loved ones or legacy contact.

- Remember privacy is more important than curiosity and information may be excluded for a very good reason. Therefore, do not ask questions about the death on social networking sites, particularly if it was a tragedy, sudden death or suicide.

- Do not share information about services or memorial events on social media without permission from the family or other primary mourners. Many people do use social media to publicize funerals if they are public events, but be sure that they are before you share or repost.

- Do not post vague comments that might cause someone to be worried about your well-being or safety.

- Avoid *comparing* your grief to others as this appears to invalidate other people's reaction to their loss and can feel competitive to those mourning.

- Do not photoshop an image of the deceased or turn their image into a meme.

- Avoid sensationalizing someone else's tragedy. Sharing grim details, posting gruesome images or recordings (e.g., the scene of the death or accident, fire or shooting), or even posting the photo of someone's gravesite can be disturbing and painful, particularly if they appear in a social media feed when someone is not expecting it.

- Do not criticize someone since it is disrespectful and inappropriate to incite controversy on a social media posting or memorial site that was created to honour someone's memory.

- Post your personal beliefs or opinions on your own social media or in a separate digital space, rather than imposing your beliefs in a personal memorial or grief space.

Distinguishing between those who preserve their own or another's digital afterlife, those that mediate the experience of others and those that receive digital afterlife of a deceased person is important when considering the idea of a good digital death. For example, a good digital death should include considering our own digital legacy and discussing it with loved ones, as well as considering the societal impact of the evolving digital dead.

THE EVOLVING DIGITAL DEAD

There appear to be a number of trends associated with the idea of the evolving dead. There seem to be four ways of considering the evolving digital dead:

- The creation of virtual humans or digital immortals

- The recognition of a digital being

- The dead as an extension of ourselves

- An evolving exploration of death through social media

The first one of these, the creation of virtual humans or digital immortals as discussed earlier in the chapter, reflects the idea that the deceased evolves or is evolved through social networking sites. The next form of the evolving dead is the idea of the digital being left in cyberspace. For example, Kasket suggests the addition of the digital to the Cartesian notion of mind and body, thus this 'digital being' is made up largely of information captured by the surveillance devices that track us in our daily lives. This kind of digital being has less to do with the decisions made by those left behind and is perhaps more related to surveillance capitalism and the collection and retention of people's data for commercial use. Zubboff argues that:

> Surveillance capitalism . . . unilaterally claims human experience as free raw material for translation into behavioural data. Although some of these data are applied to service improvement, the rest are declared as a proprietary behavioural surplus, fed into advanced manufacturing processes known as 'machine intelligence', and fabricated into prediction products that anticipate what you will do now, soon, and later. Finally, these prediction products are traded in a new kind of marketplace that I call behavioural futures markets. Surveillance capitalists have grown immensely wealthy from these trading operations, for many companies are willing to lay bets on our future behaviour.
>
> (ZUBOFF, 2019: 8)

Here we are seeing the impact of artificial intelligence and machine intelligence writ large. Yet the difficulty is that it seems unclear if this digital being is likely to be deleted or buried post-death. Indeed Kasket suggests that Facebook may be hosting the digital remains of 4.9 million people by the end of the century. She argues:

> In a 2019 UK YouGov survey, 16% of the 1,616 respondents said that they would like their profiles to remain online and visible to others after death, at least for a time (Ibbetson, 2019). A quarter of respondents, however, said that they would like all social media to be deleted entirely at the time of their deaths. This design is currently inexecutable, in both a practical and legal sense. Even if it were possible, would we seize the opportunity to make provisions

for our digital estate when two-thirds of us already fail to make plans for our physical one (Chapman, 2018)? . . . Our personal data is simply too voluminous too widely spread throughout the datasphere and too under the control of innumerable third parties to be able to simply gather it up and 'bury' it even if that's what the deceased would have wanted.

(KASKET, 2021: 21)

It is clear that there are overlaps between the digital being Kasket suggests and the creation of a digital immortal, although much of this depends on whether the data of the dead are buried or removed as Kasket would prefer, or whether they are indeed evolved into something such as a virtual persona managed by the family of the deceased.

A different configuration of the evolution of the dead is in the form of the dead becoming an extension of ourselves through the devices they leave behind. For example, Hård af Segerstad et al. (2020) argue that the device can be seen as an extension of the body and the person or even as a prosthesis. They argue that digital devices leave behind the essence of a person and this can be seen when people preserve voice messages or use their phones to text the dead. Loved ones may also carry the deceased's phone so that they feel they are also with them or play the Spotify list created by the deceased in order to feel a sense of connection with them. This suggests an evolving relationship with the dead by the use of their mobile phone to maintain a connection. The idea of the evolving dead also prompts the consideration of fake death and the choices people are making to undertake this. This is more a case of an evolving relationship with death itself and the exploration of death. This is seen in a recent, somewhat worrying study by Nikishina et al. (2020) who examined the notion of digital death and young people's attitudes to digital death in the social network of students. They discovered different forms of digital death and the ways students react to it in social networking spaces. They explain that some students staged their own death on social networks by posting images on their homepage, in order to attract public attention, to make fun of death and to reduce their own anxiety of death by 'sharing' the fear with other students. The analysis indicates that the main reason for staging death on social networks was to reduce the fear of death by creating plans and playing. Thus rather than the person evolving once dead, what is occurring here is an evolving relationship with the construction of both physical and digital death. The idea of death is being played with in order to evolve an understanding of it. However, such practices have also had serious consequences. For example, the authors note that:

> With the relation to death as ordinary entertainment content in cyberspace, a common attitude to death is also changing. The death attributes are becoming pop-attributes, funeral ceremonies games (that students play for fun), dead students seem not to be scary anymore, compassion for deceased's relatives is also decreasing.

> (NIKISHINA ET AL, 2020: 1273)

What is clear is that artificial intelligence is being used to change our relationship with the dead and the data, whether through the creation of a digital immortal or to retain the sense of a loved one. The more sinister side of the evolving dead resides in the commercialization of surveillance data and the exploration of death by faking it, which has in some cases had alarming consequences (Nikishina et al, 2020).

CONCLUSION

This chapter has examined the idea of good digital death from a range of perspectives and suggests that issues of planning, preservation, mediation, transference and taboo management are some of the central components of this. What is also evident is that the use of algorithms, the hidden artificial intelligence of social networking sites, also has sinister overtones that must not be ignored. However, it is clear that to date few people have planned for their digital death, and for those left behind, this can have devastating consequences. The impact of digital death on the bereaved, termed here 'digital bereavement', will be examined next in Chapter 7.

Death and the Liminal

INTRODUCTION

This chapter begins by exploring understandings of liminal spaces, drawing on research in this area from higher education and coupling this with perspectives of death and experiences of bereavement. The second section of the chapter examines death, liminality and afterlife, suggesting the idea of a post-death liminal tunnel. The chapter then explores spiritual and thin spaces which are seen by many as liminal spaces. It is argued that spiritual spaces are often liminal and feel other-worldly, they prompt contemplation about the possibility of worlds beyond and in particular the afterlife. The final section examines the sense of the liminal in the context of individuals anticipating death and those who have had near-death experiences.

LIMINALITY

Liminality is a term that was first coined by van Gennep (1909), who described a psychological or metaphysical subjective state of being at the threshold of two existential planes. Although the term was originally applied to rites and rituals in small human groups, it was extended to whole societies by writers such as Jaspers (1953). The idea of a liminal state is taken from ethnographic studies into rituals, for example, rites of passage such as the initiation of adolescent boys into manhood. Turner (1982) adopted the term 'liminality' (from Latin *limen*, 'boundary or threshold')

DOI: 10.1201/9781003098256-7

to characterize the transitional space/time within which the rites were conducted. Trubshaw describes liminality as follows:

> These ethnographical examples relate primarily to liminality in life cycles . . . The concept of the 'betwixt and between' liminal state then becomes easy to recognize in contemporary western culture – think, for instance, of the wedding ceremony where the 'threshold' ceremony is followed by a 'liminal' honeymoon. Think, too, of funerary ceremonies where the period from death to inhumation (or cremation) is equally 'liminal'.
>
> (TRUBSHAW, 2003)

Liminality tends to be characterized by a stripping away of old identities; it is a betwixt and between state and there is a sense of being in a period of transition and an oscillation between states. Turner later described people in a liminal state as 'a realm of pure possibility whence novel configurations of ideas and relations may arise' (1995:97). He suggested that those in liminal states were often ritually, symbolically or metaphorically removed in order not to threaten the social order. An example of the liminal being located in a ritual space is provided by Nelson Mandela's account of his Xhosa rite of passage into manhood. In practice, young men go away for The Initiation and are isolated from their families in a *Sutu* or a reed/thatched hut constructed specifically for the occasion. During this time, they are taught by elders who pass down stories of history, culture, traditions and modern-day lessons in terms of relationships, respect and family. The most significant moment of initiation is when the boy is circumcised. Mandela speaks of the ritual, where, after the circumcision ceremony in which he was declared a man, he returned to the hut: 'We were now *abakwetha*, initiates into the world of manhood. We were looked after by an *amakhankatha*, or guardian, who explained the rules we had to follow if we were to enter manhood properly' (Mandela, 1994: 33). After circumcision, the young men spend eight days undertaking rituals that include painting their faces and bodies with mud and not being allowed to eat or drink with their hands. Thus the position in which Mandela found himself, during the period of eight days after circumcision, was a liminal space; although declared a man, this was the space in which he was located before he would enter manhood properly.

Liminality also describes a sense of in-betweenness but seems to have a stronger sense of shifting identity than the concept of *metaxis*, argued for by Plato (360 BCE), since there appear to be different forms of liminality. Liminality, however, is not a polarity; it is an all-encompassing overwhelming interruption, bringing with it a sense of being in a suspended state, which can be related to grief and the bereavement process. Hearing about the death of a loved one, particularly suddenly over the phone, can leave a sense of being in a trance-like state – sometimes a dark, barely conscious state. This initial reaction to the news of death invariably launches people into a transitional moment, a sense of holding the tension between one space and another. The next section compares responses from educational research that examined experiences of liminality with those reactions to death, such as feeling lost, and moratorium status role shift and dissent.

Feeling Lost

Trafford (2008) explored PhD supervision and found a consistent sense of conceptual lostness that students experienced as if they were slipping across diverse forms of liminality. This sense of being lost and looking for something is a response to how the liminal space is entered and negotiated. Students spoke of the realization of being lost and either needing to look for something that is present or having an expectation that this sense of lostness would disappear. Here students seem to almost value doubt as a means of moving away from a liminal space. Instead of trying to eliminate the lostness, they appear to believe it is better to value it as a central principle of learning. Often following a sudden death there is a sense of feeling lost. Yet, it is not one that is easily embraced; rather, there is often a need to seize control and organize, particularly around the events of the funeral. Yet, embracing the loss and living with the liminal, and living with the realization of being in an in-between state, does seem to assist people in being able to stand on the threshold of lostness and grief and see beyond themselves.

Moratorium Status

Sibbett and Thompson (2008) suggest that in professional development, liminality is a moratorium status similar to adolescence, where different identity statuses might be experienced. A moratorium status is where delay occurs so that exploration can happen in order to develop, create and form an identity. This, the authors suggest, might be a form of liminality, since through undergoing this process and committing to the newly formed identity, 'identity achievement' occurs (p234). Identity achievement is the idea that the identity and commitment that developed in the moratorium state are consolidated. This identity achievement in bereavement can be linked to continuing bonds, the idea that the bereaved person recognizes the status of loss, of being without the loved one, but lives too with an awareness of an ongoing bond with the deceased through memories and artefacts, which in turn can result in role shift, as discussed below. However, if identity work does not take place then mimicry may occur, leading to a sense of fragmentation. In bereavement, such mimicry would occur with the pretence of managing continuing bonds, when in fact the bereaved person has remained in a stuck place; they can pretend that they are learning to live with loss when in fact they may be living with denial instead.

Role Shift

Cook-Sather and Alter (2011) explored liminality as a threshold concept between established roles at which one can linger, from which one can depart and to which one can return, such as when undergraduate students take up a liminal position between student and teacher. The study examined the students as consultants in a staff development programme. The findings demonstrated that students changed their relationships with their teachers and their responsibilities towards their own learning. The authors argue that there is a need to see student learning not as a transition from one state to another but instead as an 'unending process of dialogue and development in multiple indeterminate states, suspended between and among established institutional roles and mindsets, always unfinished, always requiring revisiting, revisioning, and re-enacting' (p51). Further, Bosetti et al. (2008) discuss liminality in relation to female professors' efforts to reconstruct their professional identities in academe. The authors located four themes, which were (1) exploring the landscape, (2) professional competence and identities, (3) competition and isolation and (4) seeking support and validation. These themes indicated that the women were seeking to reclaim both a sense of their competence and their professional identity. Through supporting one another and discussing their struggles, some women moved through liminality and made transitions into 'post-heroic' leadership. However, others going through this transition dealt with the tensions by withdrawal, working from home, or staying behind closed doors. Role shift can also be seen in the management of the bereavement process, the ongoing and unending dialogue with the deceased, the sense of being suspended between the role of a wife and a widow as well as the constant need to revise, revisit and reenact the life with loss.

Dissent

A different form of liminality was that of dissent. For example, Kofoed (2008) examined transitions in which a school pupil questions and transgresses established rules and regulations. She argues that the professionally organized setting of a school required the necessity for unambiguity to be present, and that only one way will be correct, which guides the way in which subjectifications are performed within schools. For example, there are rules for spelling, rules for mathematics and adherence to these rules is reflected in the pupils' grades and subjectification. Pupil(s) who transgress the rules become marginalized. Kofoed poignantly remarks 'Liminality is the space which stretches out between the various demands for unambiguity' (p 209). Liminality then may offer a way of conceptualizing imperfect and inconclusive transitions. Certainly, the parents of children who have committed suicide have experienced an imperfect transition in attempting to accept and make sense of death, which will be discussed in more depth in Chapter 8.

However, as is well documented, such spaces following the death of a loved one or anticipating one's own death can be stuck spaces. What seems to be apparent is that stuck spaces are dealt with in five ways, namely, to *retreat* from stuck spaces, to *postpone* dealing with it, to *temporize* and thus choose not to make a decision about how to manage it, to find some means to *avoid* it and thus create greater stuck spaces in the long term, or to *engage* with it and move to a greater or lesser sense of integration. This is exemplified in Figure 7.1, which illustrates responses to grief in relation to threshold and liminal spaces.

Responses to spaces associated with death, because of their continual construction, vary and change over time, and as Maddrell (2016) points out, since the primary space of mourning is embodied in the mourner, the embodied space of the mourning within the mourner therefore alters over time.

Retreat

In this position, the bereaved who experience stuck spaces choose not to engage with the process of managing it. Here they want to avoid

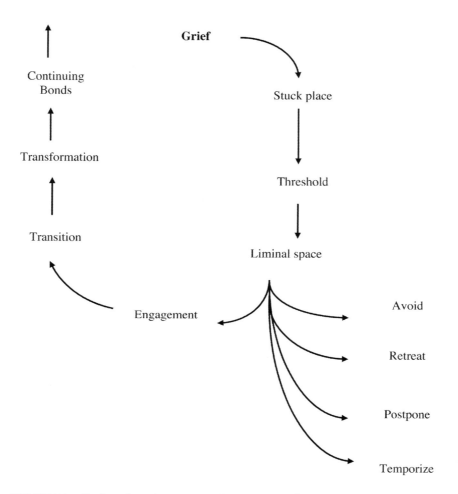

FIGURE 7.1   Cycles of stuckness towards continuing bonds.

engaging with the struggles connected with stuck spaces and often retreat behind some form of excuse, which means that they do not engage with the loss.

Postponement

The bereaved may consider staying in their position of being in stuck spaces, invariably because they have experienced stuck spaces before and learned how to manage them. Thus they choose to postpone movement. This is not a postponement of a decision, as in temporization, but a clear decision to leave life on hold.

Temporizing

The bereaved who do not directly retreat from stuck spaces may adopt an indecisive or time-serving policy. They not only acknowledge the existence of a stuck space but also have to engage with it in order to enable an effective transition to take place, but they decide that it is preferable to postpone making any decision about how to manage the loss.

Avoidance

In this situation, the bereaved do not just temporize but adopt mechanisms that will enable them to find some way of circumventing the stuck spaces. The result will be that although they may have found a means of bypassing the stuck spaces – such as being industrious, this may take more effort than engaging with loss, especially in the long term. Due to the nature of bereavement stuck spaces, they will still have to engage with such stuck spaces in their life-world in order to avoid always becoming entrenched in this position each time they are bereaved.

NB Retreat and Avoidance can be on the edge of liminality, or within liminality where a position of turning back may occur. Temporizing and Postponement tend to be in the liminal space and the bereaved may stay in these places for some time before engaging.

Engagement

Engaging with stuck spaces requires that the bereaved acknowledge their existence and attempt to deconstruct the causes of stuck spaces by examining the relationship with both their internal and external worlds. Through this reflexive examination process, the bereaved can engage with what has given rise to the stuck spaces and they are then enabled to shift towards a greater sense of integration.

There is a sense that when people are stuck they tend to retreat to places of safety, places that are known and comfortable. Such spaces are ones in which people feel they can regain control, place boundaries around their stuckness and thereby manage it in some way. As people move towards a catalyst and begin to engage with stuck spaces, they may retreat to a space of safety and thus avoid engaging with liminality and the subsequent catalyst out of liminality, thereby preventing movement over a threshold. They may hope that by retreating they will avoid moving into a liminal space, but it is more likely they will locate themselves in a cycle of stuckness.

DEATH, LIMINALITY AND AFTERLIFE

Liminality appears to be a concept and an experience that is full of different types of encounters. In relation to death and the afterlife, it seems to provide a word for the complete confusion associated with death, from questions about the purpose of death to why death has occurred. It is also used to refer to the shock of death and being in a troublesome space, as well as being used to describe the confusion associated with where the dead person might now be located. Whilst there has been much research and discussion about bereavement, questions about the afterlife are ones that remain inchoate. Further, even asking people about where the dead go can result in feelings of liminality, a sense of confusion, of being stuck and perhaps a realization that one's thinking about this is at an in-between state. Few people have discussions about where the dead go/are located and whether the afterlife itself is a liminal space; but some of the religious perspectives seem to suggest this to be a time of waiting or transition. For many people, there is a belief that the dead have sentience and presence in another space, as one priest explained to me:

> the image of the dead being in a boat travelling towards the heavenly goal, alongside the boat that we the living travel on is an image that I found very helpful. Of course, it sits uneasily alongside the idea of the dead being asleep and buried in the ground until the Last Judgment

For many people with a faith, as mentioned in Chapter 2, there is the idea of liminality, as it were, written into the idea of an afterlife. Thus, after death, there is a time of waiting or a time of transition. For example, in Christianity and Judaism, there is a time of waiting or journeying, as the believer waits for resurrection with the other faithful souls. In Buddhism generally, there is the idea of a reincarnated (*samsara*) journey of the soul. In some forms of Buddhism, texts are recited for the body of the deceased, which is thought to revive the life force within the body, so that it can embark upon the journey through the intermediate stage between death and rebirth. In Islam, deceased souls remain in their graves awaiting the resurrection on the Day of Judgement, when they begin to feel immediately a taste of their destiny to come. Those bound for hell will suffer in their graves, while those bound for heaven will be in peace until that time.

The waiting times post-death seen in many religions can be equated with the idea of a liminal tunnel, which fits better with some religions better than others. The original idea of the liminal tunnel was described by Land, Rattray and Vivian (2014) as beginning with a 'portal or gateway' triggered by the threshold concept or disjunction. Meyer and Land state:

> A threshold concept can be considered as akin to a portal, opening up a new and previously inaccessible way of thinking about something. It represents a transformed way of understanding, or interpreting, or viewing something without which the learner cannot progress.
>
> (MEYER AND LAND, 2006: 3)

The original idea of the tunnel was related to learners moving through the tunnel, through the liminal space and then emerging with a shift in conceptual, ontological or epistemological nature. Yet it would seem that post-death, there is a sense of entering the liminal tunnel, then encountering liminal spaces within the liminal tunnel which are suspended states and serve a transformative function, then leaving the tunnel to await judgement and reincarnation, and in some faiths, the development of the soul occurs in the waiting time. Different faith journeys through the liminal tunnel that occur post-death are explained in Table 7.1.

While the liminal tunnel for the dead can be seen as a space of waiting and, in some cases, movement, for the living, liminal tunnels take on a different meaning. Here, digital spaces can be seen as liminal tunnels linked to transitions in mourning journeys and the idea of digital continuing bonds and phases of transition, as represented in Table 7.2.

In the theory of continuing bonds, the bereaved learn to live with loss, still feeling the ties and creating a new relationship with the deceased. Klass (2018) argues that maintaining a connection with the dead is a 'continuing bond', a common aspect of bereavement in all current models of grief, and authors such as Kasket (2019) and Sofka (2020a) suggest that it is also evident in the use of digital media in the grieving process. What is not yet clear is whether the use of the digital results in unhealthy continuing bonds, as the digital connection may make grief more complex to manage. The idea that the deceased live on in cyberspace might be damaging to those left behind and certainly the work of Kasket (2019) and Bassett (2020) seem to suggest that the impact of digital continuance of bonds is troublesome.

TABLE 7.1 Post-Death Liminal Tunnel

	Death: moving into the tunnel	Being in the liminal space	Movement within the tunnel	Triggers to movement	Exiting the tunnel	Consequences
Buddhism	The beginning of the journey to enlightenment	Chanting texts will generate merit that can be transferred to the deceased and help them in their rebirth	Funeral rites and prayers of loved ones as well as grave goods	Festivals and grave goods Reincarnation	Final reincarnation before enlightenment	State of enlightenment
Christianity	The deceased moves from death into sleep, while waiting for the second coming of the Messiah	The soul stays in the liminal space until the Messiah returns	Development of the souls during the time of waiting but not clear how	Unknown	Christ returns and heaven comes down to earth	Souls and a new body reunited with God and with other believing loved ones
Hinduism	Death is a transition to afterlife	How lived material world affects the form of his or her next earthly incarnation	Funeral rites and prayers of loved ones	Festivals such as Diwali	Final reincarnation before enlightenment	*Moksha* when the soul is liberated
Humanism	Not applicable as no belief in afterlife of any kind					

(*Continued*)

TABLE 7.1 (CONTINUED) Post-Death Liminal Tunnel

	Death: moving into the tunnel	Being in the liminal space	Movement within the tunnel	Triggers to movement	Exiting the tunnel	Consequences
Islam	Death is the return of the soul to Allah and the transition into the afterlife	Soul is in the space between paradise and torment	Funeral rites and prayers of loved ones	Testing by two angels	Last Judgement	After Judgement, there will be a resurrection of the dead which will bring everlasting bliss to the righteous and torments to the wicked
Judaism	The dead journey towards resurrection while waiting for the first coming of the Messiah	The dead encounter experiences and angels on the journey	Development of the soul during the time of waiting but not clear how	Unknown	The arrival of the End Times	Movement into paradise or hell
Sikhism	At death, the soul is reunited with God and the body returns to the earth					Good deeds ensure that reincarnation is unlikely

TABLE 7.2 Liminality and Digital Continuing Bonds

	Digital activities	Triggers to movement	Consequences	Types of digital continuing bonds
Death of a loved one	Creation of digital memorial pages, lighting digital candles	Disjunction in the form of an ontological experience	Feeling confused, stuck and frustrated. Experience challenge to previously held beliefs	Memorialization of deceased in as many digital spaces as possible in order to maximize bonds
Being in the liminal space	Believing the deceased is still active in the digital space	Movement triggered recognition of the need to learn and shift	Learning through transition and change	Sense of deceased's presence in digital spaces. Retaining and maintaining the deceased's social media spaces
Moving towards the end of the tunnel and crossing the threshold: the shift	Recognizing that digital media can be both a help and a hindrance	Movement triggered by a sudden or gradual understanding, a stripping away of an old identity and personal transformation	An ontological shift evident in the change in self-perception and identity	Focus on memories and using digital spaces as a focus for learning to live with loss
Exiting the tunnel	Using digital media to support continuing bonds and memories in a positive way	Confidence gained through threshold shift(s)	Seeing the world afresh and valuing the continuing disjunctions and subsequent shift	Using digital memorial spaces for continuing bonds and seeing the relationship with the deceased as fluid

SPIRITUAL SPACES AND THIN SPACES

Spiritual spaces are perhaps ones that have a more ethereal quality, that have liquid edges to the territory. Maddrell (2016) suggests that it is important to provide a framework that illustrates the relationship between emotional geographies of grief, mourning and remembrance without objectifying them, whilst also making sense of the quality of space and spatial relationships shaped by grief. Spiritual spaces tend to be seen as those that have a sacred element, whether pilgrimages, churches, rivers or mountains. Spiritual spaces tend to arouse affect but also are locales within which someone can evoke a sense of the sacred through prayer or some forms of meditation and yoga. The idea of the 'sacred' is a concept that is contested between cultures and the disciplines of theology and anthropology, and, therefore, can be connected to religion, artefacts and memories.

Spiritual spaces are often in-between spaces. *Metaxis* (or metaxy) is the word used by Plato (360 BCE) to describe the condition of 'in-betweeness' that is one of the characteristics of being human. Plato applied it to spirituality, describing its location as being between the human and the divine. Whelan (2008) expands the notion of metaxis in the modern age, claiming that 'we humans are suspended on a web of polarities – the one and the many, eternity and time, freedom and fate, instinct and intellect, risk and safety, love and hate, to name but a few'. Metaxis has also been defined as the state of belonging completely and simultaneously to two different autonomous worlds (Linds, 2006). The idea of metaxis reflects the state of being that many people who are recently bereaved experience, such as feeling numb, other-worldly and confused, but it is often not a space which will feel particularly spiritual, since the sense of loss is so great. A different kind of spiritual space is a thin space.

Thin spaces are often spoken of in Celtic spirituality with particular emphasis placed on thresholds. The idea is that in a thin space the threshold allows the spiritual passing from one place to another, the idea of moving into a new realm. An example of this would be passing over the threshold of a monastery for a spiritual retreat in order to spend time away from everyday life. The threshold then becomes a spiritual space of transition where consciousness begins to shift from the ordinary to the spiritual. Often, while waiting to die or waiting for someone else to die, there is a sense of a paradoxical tension, as in R.S. Thomas' poem Kneeling; the meaning is in the waiting:

Moments of great calm,
Kneeling before an altar
Of wood in a stone church
In summer, waiting for the God
To speak; . . .
. . . Prompt me, God;
But not yet. When I speak,
Though it be you who speak
Through me, something is lost.
The meaning is in the waiting.

(THOMAS, 1993: 199)

In such spaces, in the waiting, a new reality is about to dawn, but it is not yet possible to see it or understand what existence in that new space might look like. Robertson argues:

This threshold space is one of deconstruction, while we are shedding beliefs, practises, and understanding of ourselves and our world that no longer sustain or serve us. As we shed these old ways, we stand naked, wondering what clothing we will wear next. This is a vulnerable space where we stand exposed to the raw truth of reality. We are tempted to reclothe ourselves in garments that no longer fit us, preferring the safety of what they once represented, rather than waiting patiently to receive future possibilities.

(ROBERTSON, 2020: 59)

Thin spaces also have a liminal quality, a sense of being a between space, since they capture the idea that the spiritual barriers between worlds can be broken down. An example of this is seen in the story of the *Subtle Knife* by Philip Pullman (Pullman, 1997). Here the young hero, Will Parry, uses a knife to cut between worlds and the residue after he closes it is a thin space. 'A thin space' is a term that has been used for millennia to describe a place in time where the space between heaven and earth grows thin and the sacred and the secular seem to meet.

A thin place is a term used to describe a marginal, liminal realm, beyond everyday human experience and perception, where mortals could pass into the Otherworld more readily or make contact with those in the Otherworld more willingly. In ancient folklore, Thin places were considered to be physical locations where it is easy to cross between two or more worlds.

(HEALY, 2016: 6)

The challenge for many people is that because thin spaces are often spaces of transition, they are ones where it is important to wait, to learn to be – to learn to live with the liminal.

These forms of liminality, and the relationship with death and bereavement, do not always appear to occur in particular stages or cycles. However, it seems that liminality also seems to occur in response to anticipated death and near-death experiences.

ANTICIPATED DEATH AND NEAR-DEATH EXPERIENCES

There have many discussions about the experiences of those facing death and the sense of liminality associated with it. Dennis Potter, for example, in his predeath interview with Melvyn Bragg in 1994 explained:

Below my window in Ross, when I'm working in Ross, for example, there at this season, the blossom is out in full now, there in the west early. It's a plum tree, it looks like apple blossom but it's white, and looking at it, instead of saying 'Oh that's nice blossom' . . . last week looking at it through the window when I'm writing, I see it is the whitest, frothiest, blossomest blossom that there ever could be, and I can see it. Things are both more trivial than they ever were, and more important than they ever were, and the difference between the trivial and the important doesn't seem to matter. But the nowness of everything is absolutely wondrous, and if people could see that, you know. There's no way of telling you; you have to experience it, but the glory of it, if you like, the comfort of it, the reassurance . . . not that I'm interested in reassuring people – bugger that. The fact is, if you see the present tense, boy do you see it! And boy can you celebrate it.

(BRAGG, 2007)

This sense of being present and being changed by the near presence of death is also described by Fullerton (2020), who described the liminal space she encountered following a cancer diagnosis, suggesting such a space prompts questioning about reality 'because our grasp on life has been loosened' (p78). The result of the loosening was a shift into a liminal spiritual place where brokenness became a source of growth and transformation.

In relation to near-death experiences, Murray (2010) argues that near-death experiences have resulted in a specialization of the relationship between the living and the dead. She notes that near-death experiences take place in a liminal space between the land of the living and the land of the dead, with common elements of such experiences being evident across people's stories which include:

- Moving through a tunnel, often towards bright lights

- Encounters with deceased loved ones

- A life review

- A strong sense of peace or fear

- Encounters with religious figures often related to their own beliefs such as Jesus, Buddha or celestial beings such as Angels

Such experiences also appear to trigger significant life changes and have often resulted in a loss of fear of death.

Near-death experiences bring to the fore questions about what individuals believe about life after death. For example, whether people sleep until the final resurrection, whether the dead have sentience or whether indeed near-death experiences do offer evidence for life after death at all. Yet, Murray (2010) suggests that near-death experiences appear to 'transgress and uphold the boundaries between the living and the dead' (p47). However, Foltyn (2021) argues that near-death experiences have become embraced within popular culture. She studied celebrity accounts of such experiences, arguing that these are not just experiences but for celebrities can be perceived as 'another performance'. Despite this, she suggests that such experiences (or performances) seemed to differ little from those of the unfamous and, at the same time, illustrate that many people still believe in an afterlife or indeed in heaven.

Individuals who have had near-death experiences appear to be border transgressors who interrupt notions of rightful places for the dead in society. Thus, perhaps it is the case that the body itself can be seen as a liminal space. For example, bodies that are dying can be seen as betwixt and between spaces, and Longhurst (2001: 2) has suggested the dying body appears to be a threat to spatial norms as they create and present leaky, messy and awkward zones – bodies that cannot be trusted in public spaces and which can upset domestic spaces. Thus, it would seem that dying bodies and bodies that have almost died clearly appeared to *be* liminal spaces as well as being *located in* liminal spaces.

CONCLUSION

Liminality tends to be referred to rarely in relation to understandings of afterlife and the spiritual realm; yet there does appear to be a fascination about where people are located post-death, and the result is that physical forms of memorialization are becoming an area of increasing complexity, fascination and expansion. Further, digital death spaces have also resulted in the development of memorialization due to the convergence of diverse social groups through social network sites, and also since the global pandemic, increase in online funerals and growth in the death-tech industry. Memorialization will thus be explored next in Chapter 8.

Symbols and Memorialization

INTRODUCTION

This chapter begins by examining the symbols associated with death and the way in which they are transposed into digital spaces. It then examines the relationship between semiotics and spirituality. In both the physical and digital realm, memorialization occurs in a wide variety of forms from pictures of the dead to memory boards, digital memorials and even tattoos. The second part of this chapter explores the purpose and practice of memorialization.

SYMBOLS OF DEATH AND SPIRITUALITY

There are many symbols of death in the physical world and most of these have been appropriated for use in online settings, but they often have little religious significance, even if they have some kind of perceived spiritual meaning for those adopting them. It is also evident that religious language pervades digital death spaces such as the reference to where someone might be located (heaven) as well and who they have become (angel).

Crosses and Angels

Crosses largely tend to be empty crosses, rather than a crucifix (those with a crucified Christ on it). The New Testament does not specify the crucifixion of Jesus in the shape of a cross but the theologian Barclay (1998) does suggest it was the shape of the letter T and that this was significant.

DOI: 10.1201/9781003098256-8

He argues that because the letter T is shaped exactly like the tau cross and since the Greek letter T represented the number 300, 'wherever the fathers came across the number 300 in the Old Testament they took it to be a mystical prefiguring of the cross of Christ' (p79). Crosses were used as symbols, religious or otherwise, long before the Christian era, but it is not always clear whether they were simply marks of identification or possession, or were significant for belief and worship. In terms of social media, crosses tend to be used much less than other symbols, probably because they are seen as a symbol of the finality of death, compared with angels which suggest afterlife.

Angels are assumed to be beings who watch over people, or more recently on social media, there is a view that the dead become angels in cyberspace. For example, a study by Gustavsson (2011) found that most people imagined that the deceased had become angels in the context of speaking to them on internet memorial sites. It might be that the use of language might actually be less important than its performance and social function, for example, Walter (2016) attributes the popularity of angels in part to the technological affordances and linguistic cultures of online media, suggesting 'Online, angels are the natural image for the dead, especially the dead with whom continuing bonds are desired' (p18). Further, Walter et al. (2012) argue that the online dead are always accessible and often spoken about as angels. There is a sense then from Walter et al.'s perspective that angels have some kind of digital embodiment:

> Angels are messengers, traveling from heaven to earth and back, and cyberspace is an unseen medium for the transfer of messages through unseen realms, so there may well be a resonance between how some people imagine online messaging and how they imagine angels.
>
> (WALTER ET AL, 2012: 293)

Yet from a spiritual perspective, angels are seen not as the living created in an embodied physical or digital form but rather as celestial beings. For example, in Jewish and Christian religions, they are seen by some as intermediaries between God and humanity, as well as being protectors. In Islam, angels are messengers and just like in Judaism and Christianity, angels are often represented in anthropomorphic forms combined with supernatural images, such as wings. The Qu'ran describes them as messengers with wings although it is notable that few angels in the Christian Bible have wings. In Sikhism, angels are usually mentioned as messengers of death, whereas in Hinduism and Buddhism, there are supernatural beings but these do not equate with the angels of Judaism, Islam or Christianity.

Emojis

There is a range of death emojis, such as ghosts, skulls, skull and cross-bones and the grim reaper; however, the skull emoji is also used as a warning on products to indicate poison. These emojis tend not to be used in serious death spaces but used in flippant ways to suggest something has gone awry or has broken or to invite people to Halloween events online. The use of emojis is disapproved of if sent to people who have just received a terminal illness diagnosis or when sending condolences online.

Skulls and Scythes, Clocks and Candles

Skulls, scythes and the grim reapers are often used as symbols of death in many cultures and religious traditions. The skull remains the only recognizable aspect of a person once they have died. This is captured poignantly in *Hamlet, Prince of Denmark*, where Shakespeare (1609) uses Yorick's skull as a dramatic instrument, representing the theme of *memento mori* (Latin for 'remember that you [have to] die'). Using the skull, Shakespeare portrays the futility of life and the inevitability of death symbolically. The Grim Reaper symbol comprises a skeleton and a robe: the skeleton represents the decayed body and the robe symbolizes those worn by religious people conducting funeral services. However, Kearl (2021) suggests that skulls as symbols of death have been rendered meaningless in the 21st century by the use of them on buckles, bags, watches and clothing. This commodification, he suggests, makes the human condition problematic since:

> The meaningless skull, like the invisible cemetery, belittles the presence of the dead and averts any reflection on mortality and one's humble place in the flow of generations.
>
> (Kearl, 2021: 85)

Whilst Kearl's stance does seem justified to a degree, it is probably the case that few people who choose to wear skulls in this way see them as strong symbols of death.

Clocks and candles as symbols of death frequently allude to both fragility of life and the passage of time: clocks referring to the passing of time and candles both as time passing and the idea that life will burn itself. More often in the 21st century, candles are used to remember the dead and pray for the dead in churches and online (such as the Church of England site https://www.churchofengland.org/our-faith/light-candle) and as a symbol of the hope of salvation.

Whilst symbols of death are used somewhat lightly in cyberspace, they are used by many in more spiritually related ways, yet the symbol of the hashtag has at the same time become problematic.

SEMIOTICS AND SPIRITUALITY

Social media platforms have their own, often covert rules, grammar, conventions and language. The notion of affordances remains contested for some – for example, whether users have agency or not or indeed if they are under surveillance and whether they have a truly independent choice or are steered by the affordances of the social networking sites. However, it is the users who are the ones who, over time, have begun to create conventions, reinforce language and develop semiotics. For example, the hashtag has become embedded in social media through community use, rather than the social media platforms (Bruns and Burgess, 2015). The hashtag is now used to garner 'likes' and followers, but it is also used as a rhetorical device. For example, Daer et al. (2014) suggest that hashtags have metacommunicative functions:

- Emphasizing – used to add emphasis or call attention to something in the post or something the post describes or refers to; usually expressed without judgment as a comment or reflection. Examples: #evidenceofspring; #lateafternoon

- Critiquing – used when the purpose of the post is to express judgment or a verdict regarding the object of discussion (a described experience, an image, etc.). Examples: #chefdamianisawesome; #whatishethinking

- Identifying – used to refer to the author of the post; functions to express some identifying characteristic, mood, or reflective descriptor. Examples: #ihatemyself; #diabeticinshape

- Iterating – used to express humor by referring to a well-known internet meme or happening in internet culture (or popular culture, depending). Might also be a parody. Examples: #hashtag; WhatDoesItMean (attached to image of a double rainbow)

- Rallying – functions to bring awareness or support to a cause; also could be used in marketing campaigns to gain publicity. Examples: #pitbullisnotacrime; #ASUfallwelcome

(DAER ET AL, 2014: 13)

The living continue to communicate with the dead online, sending messages as if they continue to exist in some kind of other or parallel world, along with the sense that whether there is faith or belief in the afterlife or not, there will be ultimately some kind of reunion. However, there has been relatively little discussion about the language and semiotics used in these death spaces. What is clear is that a language with a subtext of control is evident in many death spaces not only through semiotics, symbols and terminology but also in the way death in digital spaces is managed and ordered in ways that suggest how mourning and grieving should be. There are systems, software and sites that encourage people to manage their grieving and reflections in particular ways. However, there are further difficulties with the language of managing digital afterlife. For example, the 'moderating' clearly locates the control with the companies who control the sites, rather than the person who has been left the legacy control. For example, the notion of 'lurking' at online funerals or on memorial sites would seem to imply that watching is inherently bad, whilst at the same time raising questions about what counts as presence in digital spaces – and who decides.

Semiotics is the study of how signs and symbols create meaning and communicate meaning, and how they affect and even change behaviour. Digital media semiotics such as the use of the hashtag enables the tracking of grief. Thus, in digital spaces grief becomes a communal, even social, activity, as exemplified in the collective mourning of Alan Rickman and David Bowie in 2016. However, communal grieving lacks clear norms, practices and language. The result of this has been the practice of grief policing in order to manage unhelpful online behaviour. The reason that norms are not clear in large social grieving spaces is the lack of shared semiotic and language customs in public threads. Further, the huge number of transient people and posts result in them being unlikely for grief language norms to emerge, as Gach et al. remark:

> People learn how to behave on Facebook by watching others and importing norms from offline life . . . When people see a public news post and its comments in their Facebook Newsfeed, they join thousands of others in a transient community comprised of one off contributions that make it challenging for prevalent norms to emerge, let alone shape future behavior.
>
> (GACH ET AL, 2017: 9)

Grieving in social media spaces then remains problematic because there is a lack of shared language and semiotics and the norms that are imported from other spaces – such as the use of emojis and 'likes', are invariably seen as trite and tactless in grief spaces.

MEMORIALIZATION

Memorialization used to be centred around graves and graveyards, but with the commodification of symbols and the increasing commercialization of death, memorialization has become more complex. Traditional forms of memorialization include headstones, and later park benches, but in the 21st century, this includes new forms that have built on and extended old traditions such as the use of photographs, new forms of cenotaphization and memorial creation in virtual worlds.

Photographs and Presencing

The cultural shift in the use of digital photographs to display emotions, remember events and create status has resulted in them becoming a communication of life rather than a representation of events, as discussed in Chapter 3. This will be explored further in Chapter 11. Digital photographs are often used for the communication of life through such practices as presencing. Presencing in a social media context refers to the individual situating themselves in place or space (such as at a funeral) and making the occasion known to others through social media by a comment, post, photograph or tweet. Photographs have become an embedded form of presencing in mourning practices and have become an attempt to communicate grief to a wider social network.

Roadside Memorials

Roadside memorials, whilst not directly linked to the digital, do bring a focus on the dead into the every day. These shrines of flowers, toys and messages present an idealized form of the life of the (often young) deceased. This mirrors media-mourning practice whereby the dead are presented as angels with a constructed and idealized identity of the deceased.

Cenotaphization

Kellaher and Worpole (2010) use the term 'cenotaphization' to suggest that there is a dissolving of the boundary between the living and the dead. Examples include the roadside memorialization and the park bench) whereby the remains are dislocated from places of memorializing and can be seen as a third space – between the memories of the deceased at home and their body in the graveyard.

A bench symbolizes a narrative of remembrance in the form of both a physical symbol and a hidden story, both in terms of where it is located and the way in which it is positioned. Benches are used for continuing the bonds with the deceased but also leaving some kind of recognition of being left behind. In many ways, these physical artefacts can also be seen as threshold spaces, as discussed in Chapter 7, or perhaps even a liminal tunnel for the living.

Examples of cenotaphization are sometimes considered to be out of place or not in their proper place.

One such example of this is the adding of memorials on Ben Nevis, the highest mountain in Scotland, where people have been asked not to leave plaques, memorials or to build any new cairns. However, plaques are still being glued to rocks and cairns are being built, but these are now being removed. To deal with this, a memorial has been created as a *Site for Contemplation* in the form of a curved stone seat and cairn near the visitor centre in Glen Nevis. This has been designated as a place where those wishing to do so can sit and reflect in peace and quiet. It is situated close to the River Nevis and offers a view towards the mountain. People have also been asked to scatter any ashes away from the summit cairn.

Other examples of what might be termed 'cenotaphization' are given below.

Ghost Bikes

Ghost bikes are bicycles painted white, locked to a street sign, accompanied by a small plaque and placed in a particular location to raise awareness of cycling accidents. According to Walker and Lane (2011), the first recorded ghost bike was in St. Louis, Missouri, in 2003. There are currently over 630 ghost bikes that have since appeared in over 210 locations throughout the world, but there appears to be no governance associated with their installation and some have been removed by those who dislike the practice (http://ghostbikes.org/).

Underwater Reefs

This began as a project by Todd Barber and Don Brawley (https://www.eternalreefs.com/the-eternal-reefs-story/) to create a material that would replicate the natural marine environment and support coral and micro-organism development. In practice, a reef ball was created with a neutral pH content and a textured outer surface, to allow the micro-organisms to grow. In 1998, Brawley suggested he would like his ashes to be included in a reef ball and thereafter this idea developed. Eternal Reefs are now used as spaces of memorialization, which are also helping to preserve and protect the marine environment for the benefit of future generations.

Cremation Tattoos

This is where the ashes of a loved one are mixed with tattoo ink and used to create a tattoo in memory of them. The ashes of the deceased are sieved until fine dust is left behind and they are then baked for sterilization, before being mixed with the ink. Tattoos range from pictures of pets and portraits of the deceased to written inked memorials.

Death Masks

Death masks have existed since the middle ages and were made through wax or plaster casts of the deceased's face. More recently, death masks have started being created digitally. For example, Neri Oxman is a pioneer in 3D printing and constructs 3D masks by depositing polymer droplets in layers. These masks explore the transition between life and death; in terms of afterlife, perhaps the most relevant exhibited collection is *Present*, which aims to prompt discussion about the transition period between life and death (Lau, 2016).

In 2020–21, when worldwide funeral practices changed due to the COVID-19 pandemic and only small groups of people were allowed to attend funerals, a series of new rituals and memorialization developed. These included practices such as lining the roads as the hearse processed, family members taking film on smartphones and streaming the funeral, and recording the funeral from cars. There were also several incidents, in the UK at least, where funeral directors and celebrants were required to enforce quota numbers of those attending when too many people arrived.

At other more orderly funerals, mourners outside the quota stood socially distanced outside the place of worship or crematorium to witness the arrival and departure of the deceased and funeral party, and maintained due decorum during the intervening period while the funeral service took place.

However, there has been a shift from physical memorialization towards more digital practices and artefacts, and some are listed below, for example.

Social Media Mourning

Social media mourning is defined here as the use of social media to mourn in a public manner. However, there is also a new phenomenon, which I term Mass Social Media Mourning, which is defined here as the idea that we are urged to mourn something that is not our grief through social media, such as the 2017 Manchester Arena bombing, or to mourn our personal loss through social media in a highly public way. Examples of social media mourning also include bandwagon mourning and emotional rubberneckers as mentioned in Chapter 3.

Virtual Graveyards

Online memorialization has developed and grown considerably since the late 1990s. The most common virtual graveyard spaces in the early 2020s are Facebook and Twitter. The value of these spaces is that they can be accessed from anywhere, have flexible opening times and have permanence and persistence. These graveyard spaces are *about* the dead but are also spaces for grief to be expressed. The focus in these spaces is rarely negative, with no perception of hell or purgatory – only the belief that the dead are angelic or in heaven and have agency. As Kasket (2012) has noted, the dead are spoken to, and they are addressed as if they have agency: 'I know you can read this'. What has been less usual until the outbreak of COVID-19, the global pandemic in 2020 where 4,204,777 people died worldwide (as of 29 July 2021), was the rise of online funerals.

Memorialization in Virtual Worlds

Whilst there are a number of instances of memorialization in virtual worlds, the most utilized seems to be the Second Afterlife Cemetery, in the virtual world *Second Life*™. This is a site in which people can bury their avatars when they decide to leave *Second Life*™, but it is also used as a space of memorialization for family and friends after someone has died. Gibson (2019) suggests that the identity creation that occurs through avatars is part of digital life but also part of death and memorialization. She argues that memorialization in digital spaces such as *Second Life*™ offers a symbolic resolution for people after death, but perhaps more poignantly for people who have gone missing, and so memorialization in a virtual world is used when there is no resolution or departure ritual. However, one of the conundrums that the practice of memorialization in *Second Life*™ raises is what it is that is being memorialized. For example, in some virtual worlds, only the memorialization of first life is permitted, whereas, in other worlds, the avatars of those who are alive but have left the virtual world are memorialized. Gibson suggests that this is the construction of a second life as 'a life with its own story its own beginning and end' (Gibson, 2019:156) and thus introduces questions about the value of memorializing an avatar when its owner is still very much alive.

Tellership

Tellership is either where those who have lost loved ones use their social media profiles to share their personal story of loss or someone who is dying broadcasts their story. An important part of tellership is the role of the co-teller who responds to the broadcast and later offers tributes, thus creating a memorial, whether they knew them or not (Giaxoglou, 2020). An example is of Scott Simon, the National Public Radio reporter who live-tweeted his mother's death during her last days and hours. Cann (2014) suggests that Simon's use of Twitter marked three cultural shifts in the use of social media in that it sanctioned the use of social media for documenting difficult events, social media was used to verify the importance of the event and it provided updates in real time that made death a spectacle.

Deathtainment

This is the mediatization of death announcements produced for popular entertainment. This activity is seen as part of the new forms of death cults, such as grief policing, media mourning and bandwagon mourning, which illustrate the increasing interest in the portrayal of death in media spaces, as discussed in Chapter 3.

Hyper-mourning

Hyper-mourning is defined here (following Giaxoglou, 2020) as mourning through multiple media sources which also includes new forms of mourning, often perceived to be overreactions to public deaths and public death events, such as celebrity mourning and global death. For example, after the death of David Bowie in 2016, the music and fashion industries both shared their grief on social media using images such as the thunderbolt, the signature sign of Bowie. A more recent example was the 2020 COVID-19 pandemic, when social media became not only commentators but also the mediators of management of the crisis in the later stages, where governments (particularly the UK) were misrepresenting death figures and cultivating a culture of fear whilst the then US President, Donald Trump, flouted global guidelines in public.

Giaxoglou (2020) suggests hyper-mourning practices relate to the type of loss which affects the way grief is shared, the intended purposes of memorialization, the duration of the mourning activity and the degree and types of activity on the site on which the memorial is hosted. She argues hyper-mourning can be viewed along a spectrum of five typical categories namely participatory, motivational, connective, cosmopolitan and rebellious, as shown in Figure 8.1.

These different types of hyper-mourning do not have distinct boundaries and tend to overlap; they are defined as follows: participatory hyper-mourning refers to memorials created by bereaved groups, such as family and friends, and are designed to be lasting memorials to celebrate the deceased's life. Motivational hyper-mourning is more akin to a living funeral or anticipatory grief whereby social media is used to document the near deceased's life and illness to mobilize grief as an inspirational force for life, such as the use of cancer blogs and vlogs. Third, connective forms of hyper-mourning can include bandwagon mourning but more often is used to share immediate emotional reactions to death news, epitomized by hashtag mourning following terrorist attacks. The next form is that of cosmopolitan mourning, where people share reactions to public and often iconic deaths and disseminate images, in some cases seeking to promote

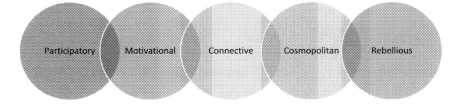

FIGURE 8.1   Typology of hyper-mourning.

a sense of global outrage. The final form of hyper-mourning is the idea of creating a particular moment – or in some cases, a movement, prompted by a death. For example, social media is used to draw attention to wrongs and thus mobilize people around political action, such as the Black Lives Matter movement.

Virtual Veneration

This is the process of memorializing people through avatars in online games and 3D virtual worlds. An example of this is ancestor veneration avatars (Bainbridge, 2017) that are used as a medium for memorializing the dead and exploiting their wisdom by experiencing a virtual world as they might have done so. In practice, Bainbridge created a number of ancestor veneration avatars that were based on historical figures. These included three deceased religious innovators in three historically grounded virtual worlds which were *Dark Age of Camelot* based on the 6th-century Northwestern Europe, second, *The Secret World* based on a historically based New England, Egypt and Transylvania, and third, *A Tale in the Desert*, which simulated ancient Egypt. Within these worlds, Bainbridge employed different avatars to explore different historical events and practices. The historical figures were used in these role-playing games to enable the players to gain new perspectives about the meaning of the world, of society, of ethics and of the mortal self. The idea of such creation, Bainbridge suggests, begins to help us to see what digital immortality may look like and such work can help us to consider the value of digital immortals.

Virtual Pages of Remembrance

Pages and spaces of remembrance are used on Facebook, Pinterest and Instagram. Much of this involves posting a picture of the deceased, often on anniversaries of particular events important to the deceased. Others post pictures of places or memories of events. Very few people, however, tend to post pictures of open coffins, gravesites or selfies from the funeral.

Digital Memorialization Following Suicide

Bell and Bailley (2017) examined the perspectives of people who used Facebook to memorialize loved ones who had died by suicide and highlighted the benefits and challenges of Facebook use in the aftermath of a suicide. It meant negotiating particular pitfalls:

tailing off interest due to time lapse, feeling responsible for the thoughts and actions of others, censorship, unwanted interest from strangers, changes to the site without their consent, removal of the page without their consent.

(BELL AND BAILEY, 2017: 83)

A further pitfall was the possibility for suicide contagion – whereby one suicide may prompt the occurrence of another. Yet, they do note that for many people Facebook did offer a means of dealing with stigma and alleviating disenfranchised grief.

Continuing and Communal Bonds

Kasket (2012) suggest that an extension of the continuing bonds theory (Klass et al, 1996) should be that of communal bonds. In her study, she found that participants found that visiting Facebook and sharing grief was more helpful for many people than visiting a physical graveside or other forms of memorial. Communal bonds are the continual posting on the person's wall, mostly to the dead person but also to the associated community of mourners. Whilst this is now common practice, it is not clear if this is helpful in the grieving process, or whether it results in those left behind being stuck in a cycle of communal grief sharing. However, in many ways, it could be likened to the sense of journeying with the dead who have gone before. Examples of this include the idea of the Communion of Saints in the Church of England whereby the dead continue on and are active in heaven (although this is contested and debated by some) or the idea that the dead live with their ancestors, as believed in many African Traditional Religions, and perhaps exemplified by the Akamba people or Nyakyusa people.

> ...the Akamba people in eastern Kenya speak of death as joining the 'company of one's grandfathers', the venerable forebears populating one's genealogical tree (Mbiti 1990: 152). By the same token, the Nyakyusa people of Tanzania and Malawi comfort the grieving by assuring them that their relative 'has gone to his fathers' (Mwakilema 1997: 20). It is the same for many traditional peoples of Africa.

(COOK, 2007: 666)

When considering the old, new and merged practices of memorialization, what we are perhaps seeing here is the integration and overlapping of persistence and permanence. Persistence captures the idea that the dead persist in cyberspace. In practice, this means that the living feel as if the dead are still there and that the digital self persists after death. Persistence here is about the persistence of the persona, rather than their artefacts. Permanence is the idea that the artefact, the profiles and the pictures of the deceased live on in media and ultimately, unlike physical photographs, cannot permanently be deleted. Permanence then means that people's artefacts live on in cyberspace; it is important to be clear that however much is deleted, something will still be there.

CONCLUSION

This chapter has considered a wide range of practices connected with memorialization. It is clear that many of the old and physical practices have been developed or have moved into digital spaces. Although many death rituals in the 21st Century mirror traditional ones, such as funerals, burials, gravestones and remembrances artefacts, the digital has resulted in the creation of broader mourning spaces and places, as well as predeath-related practices. Yet it remains unclear if these new rituals help engagement with the recognition and acceptance of death, or just result in death becoming another social media norm. Yet at the same time memorialization in the digital age is becoming more common, commercialized and mostly sophisticated due to the increase in digital artefacts, death tech companies, and diverse and accessible apps and social media being available. Despite it becoming increasingly common to memorialize in digital spaces, few people plan for the creation or ongoing persistence of their loved ones. Furthermore, as it will be made clear in the next chapter, religious experts remain unclear as to how digital media and digital afterlife creation affect understandings of death and the afterlife within religious contexts.

CHAPTER 9

Perspectives on Digital Afterlife

INTRODUCTION

It is evident that relationships between technologies and the actual content of belief and practice have received relatively little attention and this chapter begins by exploring some of the recent debates in this area. Whilst the disciplines of death studies and digital afterlife do not largely examine the spiritual side of bereavement there do appear to be increasing concerns about the impact of the digital on the bereaved. This chapter begins by reflecting on technology, belief and practice and then presents the findings from a study that examined how digital media and digital afterlife creation have and are affecting understandings of death and the afterlife within religious contexts. The final section of the chapter suggests that we are currently experiencing a shift from the digital to the postdigital and this is resulting in postdigital theologies.

TECHNOLOGY, BELIEF AND PRACTICE

A number of models that relate to the relation between belief and technology are starting to emerge, but each of these depends on the theological or philosophical framework within which the debate is set. Such debates are important in examining religious experts' stances towards digital afterlife. Reader and Savin-Baden (2021) have suggested four types but recognize that the extremes in which the digital is either coterminous with God and Creation or else totally separate and distinct from God and Creation are

DOI: 10.1201/9781003098256-9

both inadequate in addressing the theoretical and practical concerns. The four main types are:

The digital is internal to Creation/God.

The digital is external to Creation/God.

The digital is internal to the Human (embedded, embodied, etc.).

The digital is external to the Human (instrumental only).

The first approach allows no space for critical engagement, as it merely subsumes the digital into narrow parameters within a limited sphere of belief. The final approach fails to recognize the already entangled nature of the relationships between humans and non-humans, treating technology as purely instrumental. It is not yet clear from these ideal types which ones may be the most useful for exploring how theology and digital technologies are now shaping and affecting one another. One such approach would be to argue for postdigital theologies, which will be discussed later in the chapter. Other alternative approaches might move towards a more nuanced and differentiated interpretation of these relationships but still operate within quite traditional theological frameworks which risk predetermining the possibility for fluid outcomes.

As mentioned in Chapter 4, there is a range of perspectives about digital afterlife and theology and Hutchings (2019) suggests that there are three conflicting arguments: (1) Digital afterlife is compatible with religion. (2) Digital afterlife is not religious at all. (3) Digital afterlife contributes to a new kind of religion. Worthy and important though these may be, there is still little discussion about how technology affects faith itself. It is not clear whether certain religious practices provide alternative approaches to ways in which the digital is developed and deployed and whether these might contribute towards a public theology which could challenge areas where human well-being is under threat.

During the COVID-19 pandemic while churches have been inaccessible, there is no doubt that using digital technology to conduct and share worship has been a means of sustaining the life of worshipping communities. The conducting of worship services and other meetings via Zoom, Skype and live streaming have been the main ways of keeping people in touch with each other and maintaining at least some pattern of worship. The digital has enabled acts of worship and some degree of support, albeit remotely rather than face-to-face. However, it is not clear what has been gained and what was lost as a result of this and what sort of God now lurks behind the technology. This is a classic example of how the digital now impinges upon and actively influences religious practice and belief. One argument is that the technology is a neutral tool which can be deployed without any danger of either distorting or enhancing the worship and pastoral relationships, and indeed there is also no suggestion that these neutral tools have an impact upon belief. The opposite is the case, which is why there is much debate about both within religious circles. The next section presents a wide range of perspectives about the relationship between the digital and religious views and practices about afterlife.

PERSPECTIVES ON DIGITAL AFTERLIFE

This section presents the findings of a small-scale study. The study used narrative inquiry in order to understand the religious perspectives of experts on digital afterlife creation and management. It used narrative to construct data with the participants so that narrative becomes a process of meaning-making, particularly when encountering unusual events or issues. In practice, I undertook interviews with clergy, public intellectuals, academics and a bereavement counsellor. The findings of the study suggest that social media could be changing perspectives on understandings of digital afterlife, memorialization and views about death and theology. The aims of the study were to:

- Understand the impact of digital media on those living and those working with the bereaved: religious leaders, researchers and academic experts.

- Promote discussion about digital memorialization post-death.

- Understand the impact digital afterlife creation may have on conceptions and theologies of death.

Central to the use of this methodology were:

Bias – this was acknowledged by the researcher and the researched and together we sought to understand the impact of the way stories collected over time changed perspectives, views, bias and the stories themselves.

Ethics – Ethical clearance was gained through the University of Worcester Ethics Committee, with the agreement that the interview process and questions used would evolve through the study.

Informed consent – the nature of consent was deemed to be something that would be discussed and negotiated as the research progressed. We agreed that the signing of the consent forms would be undertaken after the data interpretation had been mutually agreed upon, since consent can never be something that is fully informed, as participants rarely completely understand what they have signed up for or indeed what is published as a result of what they have said, even if they have signed a form stating that they do so (Savin-Baden and Major, 2013).

Reflexivity – In this study, reflexivity was a recursive process so that interviews were both reflective and reflexive, in short, they were reflective transition spaces. Transitional reflective spaces (Savin-Baden, 2008) are encountered when a challenge or query prompts us to reconsider views and perspectives. However, there was also a recognition of the degree of reflexivity occurring through the interviews for the researcher and participants.

Data Collection – The Interviews

Bruner stated that to narrate derives from both 'telling' (*narrare*) and 'knowing in some particular way' (*gnarus*); thus the two are entangled (Bruner, 2002: 27). Narrative requires recounting events to construct *with* the participant the sense of the issues under study. Bruner (1990) also believed that narrative is a process of meaning-making, particularly when encountering unusual events or issues. Thus the interviews took place as conversations, discovering views and exploring ideas. In total, 12 in-depth interviews and 4 email discussions took place with Ministers, Counsellors, Public Intellectuals, Technology Researchers, Academics and Business leaders who were Buddhists Christians, Hindus, Humanists, Jews, Muslims and Sikhs.

Analysis and Interpretation

The recordings were transcribed verbatim and transcriptions were analyzed narratively to produce the most authentic and in-depth portrayal of each experience. The narrative inquiry focuses on the contextual dimensions within the transcripts by analyzing contiguously and making connections among personal and social interaction, past, present and future and the notion of place (Clandinin and Connolly, 2000: 50). Finally, the completed stories were analyzed using a constant comparison approach where textual data is unitized into categories, assigned code names and then collapsed and expanded until themes emerge to provide the commonalities across these different accounts.

The findings presented here and throughout the book, indicate a diverse set of views from participants that could have been expected to have more similar perspectives given their employment. Whilst the themes across the whole data ranged from ethics to persistence, just four of these are presented here, namely death meanings and practices, memorialization, theology and persistence, and perspectives on digital afterlife.

Death Meanings and Practices

Within this theme, two key issues were raised; one was the denial of death and the second was media as an unwelcome reminder of death. The participants spoke about how most people in society ignore the presence of death. This they suggested is seen in the lack of planning for death whether it be the lack of a will, little or no financial planning for the funeral or little preparation for the consequences of the loss of a loved one For example Paul and Carolyn explained

Carolyn: I think most people are so unaware of planning for their own death on any level, let alone digital. We see it as a huge problem generally; people simply do not think about all the things that need to be done when someone dies before it happens. Let alone thinking about getting ready for any sort of digital memorialization, they haven't even thought about physically how to pay for a funeral.

Paul: For most of the population there isn't one *(a theology of death)*. They don't know what death means still. You'd think that after all the practice that the human race has had at dealing with it, we'd deal with it better. But we don't. We've not really adjusted to the fact that we're going to die.

In terms of social media, participants spoke of it being both a welcome and unwelcome reminder that someone had died. For some people, social media was a place where people felt they could leave messages for the dead which they believed helped the grieving process. On the other hand, the sense of the dead lurking in phones and living in cyberspace was also something that was seen to be both uncomfortable and troublesome, as Andy explained:

> I have seen digital media play the role in what I would call, communal bereavement. Where say the Facebook profile of someone becomes the place where their friends and family can remember them. Or can choose to leave messages for them, as if they're still here, and that's a real part of the grief process. But I've also seen times where social media is kind of an unwelcome reminder and people really don't want to be reminded that their family member or loved one is no longer with them.

This ambivalence about the location of the dead being troublesome was a common view across all participants, but it also related to the issue of memorialization.

Memorialization

Participants spoke about the culture associated with memorialization, as being associated with the need to care; Esther, for example, spoke of the sense of wanting to manage the presence and visibility of the dead in cyberspace. Anna mentioned this in a similar way, suggesting that digital spaces create new spaces and communities for the bereaved. The digital community of bereavement was used to share memories and stories as well as to care for the bereaved. As well as this idea of a culture of care for the dead, there was also the awareness of remembrance needing to be tasteful, not gaudy, which was also related to culture. For example, Harry explained that putting ashes in bullets is largely more common in the US, whereas benches are more common in the UK. However, he noted the issue of temporality related to memorialization:

> There was a gypsy thing that got into the news because it was too gaudy. So it's making the point that this kind of wide diversity about how people get remembered, and people's taste and culture is all part of it . . . such as a bench in Kew Gardens – and money comes into it. It costs I think 5k to have that *(a name)* on a bench in Kew Gardens, but then also about temporality; you get ten years and then you pay again. So this idea of immortality or enduringness is fake.

The suggestion that enduringness is fake was spoken about by Harry in relation to a physical memorialization, but this would also seem to be the case in digital spaces such as online Memorial Gardens and on Facebook memorial pages.

Other memorialization practices were spoken of in terms of the way these had changed with the advent of the digital and examples were cited by participants of Facebook memorial pages and multimedia funerals. Maria felt that online funerals sanitized death, so that the language used both in relation to online funerals and on social networking sites seemed to trivialize death. She explained:

> It *(an online funeral)* doesn't feel the same. It doesn't feel like it's got the same weight of depth of the ability to actually be really bodily in that moment and really engaging with what you're feeling, experiencing, thinking, and all the messiness of that, because it feels sometimes in the digital world it can be a bit too tidy . . . it almost strips out the idiosyncrasy of response, and I think it's where those kind of catch phrases almost become so debased in a way. The language of love and love and prayers. Is it love and prayers that always gets used, love and prayers. In the States, I think particularly after shooting incidents, there'll often be this thing where they're always saying love and prayers, or thoughts and prayers. I think it's something like that. And it feels like it's almost a sanitised way of engaging with something.

Heather's perspective overlapped with Maria's in that she spoke of the shift in online memorialization as a form of transposition:

> It's almost like we're transposing a lot of our more sacramental, more sentimental ways of hanging onto a memory of someone into the digital world, aren't we? It's almost like the virus that we're all worried about now has done a species hop, hasn't it? So it's jumped and causing us trouble. And it's almost like our attitudes and our habits have made a jump into the digital in the same way

Yet for both Heather and other participants, this transposition into the digital was troublesome; the idea that our attitude to the dead in cyberspace had shifted without our realizing it. For many of the participants, there was a discomfort with this transposition, but few were able to really voice what it was that they found so disruptive, and it was perhaps the way in which the persistence and permanence of the dead had become more fluid in cyberspace than in the physical world. Yet, Alastair suggested the digital memorialization was actually positive; that it was an extension of love:

> . . . to say that they live on digitally is a kind of step up of them from saying that they live on in memories, and there's just a place where they are . . . But the thing it reminded me of was that thing from Harry Potter where Dumbledore could take all his memories out and put them in something, empty his head because it got too full, so he needed to take, drain his brain as it were, and put it somewhere. And in a way, that's what you're doing with digital memorialization, isn't it? You're saying let's take all this bit and actually preserve it somewhere, where it will be safe. And that's living on in love, isn't it, really?

For Alastair, memorialization was not just an effective way of preserving someone and their memories, but it was an opportunity to ensure that the love and the memories of the deceased remained part of the love of the living.

Theology and Persistence

Participants were largely unsure about the impact of the digital on theology and eschatology. Carolyn wondered if the digital would challenge the way theologies of death were seen. Others such as Maria reflected that the digital might be resulting in reducing people's fear of death, she explained:

> I think one of the things I find kind of fascinates me is the extent to which when somebody does die, there is a retreat possibly into that language, so that it's still there, really. I'm wondering if people hope that the digital might provide the solution to the fears of death. I'm not so sure. I'm just not convinced, really. I don't know . . . Are religious institutions actually actively looking to see what they can do with this, or is it really something that is about a kind of post-secular how do we still hold on to some sense that there might be a life after death for us? I don't know.

Like Maria, Anna was interested in the impact of the digital on the bereaved, suggesting it was changing bereavement behaviours:

> The other question was about whether I thought digital media was changing theology of death. I'm trying to think of a way in which it might be. It's probably changing bereavement. Whether it helps people to grieve or holds people back so that we don't let them go, I don't know. People did argue that having photographs captured people and it meant that you held them, if you like. If you had a photograph of somebody, it kept them with you when you need to go. I wonder what the Facebook interactions does for people in terms of the bereavement process.

The idea of holding people in photographs, as Anna mentioned, was linked to the idea of being able to hold people, to track and trace people and to 'retain' them in the digital. There were two competing perspectives about the idea of retaining people in the digital. Heather argued against deletion and Esther's concerns about the impact on both the environment and the bereaved led her to argue strongly for deleting the dead:

Heather: I think it's exactly the same as me deleting that other friend from my phone. I would feel it was killing them again. I mean, you can't stop people dying, but oh my goodness, you don't have to go in and kill their memory. That's weird, isn't it? But I think it would feel like killing them again. Legacy is incredibly important.

Esther: But deletion upon death is not necessarily horrible. I was really upset when Twitter cancelled its cull in November. So, in November of 2019, it announces an imminent cull of inactive accounts that was due to happen mid-December. Twitter has always had an inactive account cull policy. It's just that it's never been consistently enforced. And they have never said out in the media, we are going to do this. And there was an outcry from bereaved people. And one day later, Twitter backpedalled and said, we're not going to do the cull. We're going to look into a memorialization, like our friends over at Facebook and Instagram. And I just slapped my forehead in despair.

Digital Afterlife: Understandings and Perspectives

Understandings about digital afterlife were spoken about by many participants as being created by default through social media, which brought with it almost a sense of death and afterlife becoming trivialized. Yet, at the same time, there seemed to be a feeling of detachment that the creation by default or by design was somehow separate from the spiritual realm and the idea of where the soul would be located physically after death. For example, Maria suggested that technology may result in the trivializing of afterlife, because of the ways in which death becomes reshaped by the use of technology, resulting in a lack of acceptance of the limit of life. Yet, at the same time, there was a belief in the sentience of the dead in social media, as Esther argued:

> Many, many people that I have spoken to without formal religious beliefs experience the presence or the interaction or the positioning of ongoing digital selves on the internet as a kind of afterlife. And furthermore, an afterlife with a kind of sentience somehow. With a feeling of still being in some kind of social contract with the dead person, which you can see in all sorts of ways. But the sort of sense of obligation about going on at certain times or saying things at certain times. Or reaching out at certain times, like anniversaries. Whether death anniversaries or birth anniversaries or significant events . . . And of those occasions, when I have dug into it with participants, of whether they believe the person is actually in receiving the communications, even those people who aren't particularly religious or spiritual, they do talk about this felt sense that they can't really explain. They don't really know why they feel that or believe that, but they just do. And I think that makes a whole lot of sense, particularly for people who have grown up with digital technologies.

Esther's views are based on years of research into people's perspectives of death in cyberspace. Her stance and her experience illustrate that both belief in the sentience of the dead and the possibility of a social contract with the dead were common for many people, whether they had a religious faith or not. This conviction in the ability to communicate with the dead, whilst not new, does differ from practices such as seances, where an intermediary is used to communicate with the dead. What seemed to differ was that there was a belief that the dead were close, near at hand in cyberspace, almost as if they could hear and respond to the living. Yet, Alastair offered a different stance, as he reflected on whether sentience would mean that the avatar could feel pain, suggesting that perhaps we imagine an avatar would not feel pain, in a similar way to the expectation of no pain in paradise:

> If we have sentience, does that mean we will feel pain? And we imagine not. We imagine paradise is free of pain. And yet, God is not free of pain. God chooses to take on pain, because love always feels pain.

Thus assumptions about the afterlife linked with the way in which death was perceived and the practices related to this. Most of the participants in the study found the possibility of creating a post-death avatar troublesome, and there were a wide variety of perspectives on the value and impact of digital afterlife, which included issues such as risk, responsibility and agency. For example, there were many debates in the interviews about the impact of technology on the environment, in particular artificial intelligence. For Esther there were two main concerns; the first was the ethics of sentient AIs and the second was the legal and ethical concerns associated with digital afterlife creation:

> If we start creating all sorts of avatars that allow for a kind of continuity or persistence of somebody that once had a legal personality with financial assets and various kinds of interests and all sorts of accounts all over the web . . . So, I think there is already a problem there. I think there is a huge potential for identity theft, with respect to accounts that are not closed down. So, email accounts, social media accounts, bank accounts. Other things that aren't closed down, partly because, again, if people aren't making provisions, or there aren't systems to make provisions for how all of these things can be notified by the estate, then there will be long periods of time where things are left open. And enterprising criminals can combine. They can steal a person's identity during that period after death and they can exploit all of these unclosed down accounts. You introduce avatars into the mix that can approximate that person, it's a little bit more dangerous.

Esther was concerned with not only the use of avatars for memorialization and the way it would affect the ability to grieve but perhaps more importantly the additional issue of identity theft. Whereas for Maria, the concern was that digital afterlife creation would result in an amenable obedient copy of a loved one:

> I think the thing that always strikes with human relationships is that we invariably misunderstand each other. Often it's quite messy and difficult and you never really know . . . Well, you don't know. You don't know what's going on within inside somebody, in relation to how they are with you or think about you . . . So my concern, possibly, would be perhaps my fear of it being a little bit like creating a kind of Stepford Wife who is going to be amenable to our desires in a way that in real life they wouldn't have been, because they would've had their own understanding of things, their own red lines, their own sense of irritation with you or whatever. In a way, those are the things that actually make our relationships most rich and most important to us. Whereas if it's just something that's effectively a kind of automaton that is modelling how we would like them to be in relation to us, then I don't really see that that would be a good thing, actually, for our own need, which I do think is a need, to descend to our own obsessions and desires and all of the rest of it.

Such a creation could not only affect grieving but also result in overdependence on a being with no responsibility and no sentience. Anna had similar concerns and raised the question of who would be responsible for the afterlife creation:

> My query about having an ongoing immortal avatar is, who holds that, who is responsible for that person and what that person is saying and doing, because it seems to me that there are two options . . . It seems that you can either create the avatar and it's you and remains as you when you were living, and people can engage with that, or there's a possibility that it can then have its own existence beyond. Am I right, that's right, isn't it, because it can still become. That's my question, is who is responsible for that? I imagine you set up the algorithm so that the avatar has its own life, but I wonder where the, as a Christian, the soul of that

person is, where is the person, really. But in an ethical stance, I would wonder who had responsibility. If that avatar was doing things that were illegal or immoral, who is responsible for it? Also, if that avatar is thought of in any way, shape or form as a person, what happens if the technology either fails or changes? Can you pull the plug on it? Can you switch that person off? In which case is it murder or suicide if the avatar is switched off? There are ethical questions, but there are also Christian questions, I suppose, about whether that person has a soul.

Whilst there are many death tech sites and increasing options for creating a digital afterlife, there remains little legal or ethical guidance in this area, which will be discussed in more depth in the next chapter, Chapter 10. However, Harry offered a different view.

If you imagine two people dying in 1970. One widow might have lots of photos of her husband, and the other widow might have no, or hardly any photos. So in that situation, if you take that thought experiment, which widow would you rather be? I think I'd always want to be the one who had access to photos. Or maybe that huge book of colour photos would make it harder for me to let go, but I'd rather have the possibility. Do you know what I mean? . . . So to me, the same argument would apply to two people losing their partner in 2025. I think you'd want to have access to information, and then be empowered to decide to figure out how you cope with it, or how you let go of it. So in other words, I suppose what I'm saying is I think with tech I believe in agency and people's ability to have power over tech basically. We may get that it shapes us in the McLuhan sense, but things like this; it's a tool, it's a resource and people can use it.

Harry's stance of seeing the tech is useful, and something over which we have agency is positive, but it is a stance that tends to ignore the problem of affordances; affordances captures the idea that we are encouraged to use technology in particular ways and to operate within particular sites online because of the way that they have been designed to encourage particular usage and ways of operating. Certainly, Reader (2021) suggests that agency should also include the non-human, in this case, the digital, and that the latter shapes the human as much as the human shapes the digital.

The complexities around the debate concerning affordances also introduced questions about the location of the soul in cyberspace from Anna and Harry, as well as the position of the soul in afterlife creation; Carolyn reflected:

> Apart from watching Years and Years, *(TV series)*, last year; that was amazing, that drama. Apart from that, myself, it was a kind of startling move at the end when I think she was becoming something else wasn't she and living in the Alexa device or something. It wasn't quite clear; something as extreme as that? . . . , at the one extreme, which is almost like the . . . Is it the halogenic thing of getting frozen, and then they hope that technology in the future will bring you back to life.

It was evident from the perspectives of the experts interviewed that it is unclear how theologies and digital technologies are intersecting, particularly in complex areas such as authority, authenticity and bereavement.

Across the data, there was a sense of shifts and changes occurring in religious beliefs and practices that were often unobtrusive. For example, there seemed to be a transposition of sacred practices into the digital and thus changing practices. Transposition occurred in, for example, the ways in which funerals were performed, such as the transposition of power from the celebrant to the funeral director managing cameras and voice-overs, almost operating as a chat show host interrupting mourning at the graveside. In other instances, it was possible to see the migration of the sacred into the digital, as it were creeping in to spaces unnoticed. This was seen not only in the use of religious language in social media spaces and the assumption the dead live on as angels in cyberspace but also in the perspective that religious cultures were being changed subtly by the digital. However, a different perspective could be that the digital is shifting the sacred into a thin space, making the sacred more accessible in online places of worship with the unnoticed spirit in the digital, working behind the scenes, and prompting a shift too, towards postdigital theologies.

THE ARRIVAL OF POSTDIGITAL THEOLOGIES?

What many of the perspectives in this chapter illustrate is that we are not just operating in a digital world but to a large extent a postdigital one. There are many diverse understandings, definitions and stances about the term 'postdigital'. For many authors 'postdigital' is not temporal, it is not 'after' digital, rather it is a critical inquiry into the state of the digital world. Cramer (2015) provides a complex list of characteristics of the postdigital which are summarized in part below, along with the work of Jandrić et al. (2018), Jandrić (2019), Peters et al. (2021a, b), Savin-Baden (2021b), Andersen et al. (2014) and others. The postdigital is seen as a stance which merges the old and the new; it is not seen as an event or temporal position; rather, it is a critical perspective, a philosophy, that can be summarized as a collection of stances in the following intersecting positions, as adapted from Savin-Baden (2021b):

- A disenchantment with current information systems and media and a period in which our fascination with these systems has become historical (Cramer, 2015).

- The merging of the old and new by reusing and re-investigating analogue technology with a tendency to focus on the experiential rather than the conceptual (Andersen et al, 2014).

- A continuation of digital cultures that is both still digital and beyond digital.

- A mutation into new and different power structures that are pervasive, covert and, therefore, less obvious than other power structures.

- A blurred and messy relationship between the analogue and the digital, humanism and posthumanism, and physics and biology (Jandrić et al, 2018).

- The emergence of technoscience systems, which are transforming and transformed by global quantum computing, deep learning complexity science, artificial intelligence and big data (Jandrić, 2019).

- The emergence of postdigital humans: their conceptions, philosophy, ethics and religion (Burden and Savin-Baden, 2019; Savin-Baden, 2021b).

- The involvement of many connected questions such as digital immortality (Savin-Baden and Mason-Robbie, 2020).

- The condition of the world after computerization, global networking and the development and expansion of the digital market.

The postdigital then is not just about positions or spaces inhabited just for a time, it is essentially ungraspable. This ungraspability relates to the way in which structures, political systems, cultures, languages and technologies differ and change. The digital world already exists and determines to different degrees our biological, cultural and political lives. Derrida suggests that, by deploying digital technology, religion(s) risk betraying the very beliefs and practices that are fundamental to their existence. Derrida describes this as autoimmunity – destroying the means by which one protects the values to which one claims to adhere. Derrida employs the concept of autoimmunity to address the relationship between religion and what he refers to as technoscience. One of the characteristics of religions is that they point towards the infinite, that which is of ultimate value and should remain safe, sacred, holy and set apart. When they become articulated through the automated mechanisms of technology, this runs the risk of undermining their own beliefs and practices (Derrida and Vattimo, 1998: 50–51). Such a conundrum was evident in the perspectives of the experts interviewed here.

CONCLUSION

This chapter has drawn on the findings of a study that examined experts' perspectives on digital afterlife. It is evident that this is an area that many still believe is underdeveloped and troublesome and that it is unclear as to what extent digital media are complementing or replacing well-established formal structures and religious rituals, as well as beginning to create formations of local theology in digital spaces. Furthermore, the perspectives presented in this chapter also illustrate that there remain questions about the legal and ethical consequences of digital afterlife creation on relatives, friends and societies, which will be explored next in Chapter 10.

Digital Legacy

INTRODUCTION

The digital world introduces a diverse range of legal complexities about how digital assets should be managed. This chapter will explore legal issues and ethical concerns in relation to digital afterlife and the spiritual realm. There is currently little understanding of exactly how digital ethical concerns in relation to the dead are being dealt with at both practical and policy levels and legal conundrums remain troublesome. This chapter begins by presenting the recent research and literature on digital legacy, then explores issues of ownership and privacy. The next section examines digital estate planning and the death-tech industry and the chapter concludes by exploring the complex area of ethical responsibility.

DIGITAL LEGACY

Digital legacy encompasses all the digital property left behind by the deceased, and these include digital remains, digital assets and digital traces, as well as digital endurance, as illustrated in Table 10.1.

Digital Remains

The concept of digital remains is used to describe the remains of the dead on social media sites and digital spaces such as photographs, websites and memes. Stokes (2015), in coining this term to describe the data of the dead, suggests that seeing them as remains is raising the status of the data to that of a corpse.

DOI: 10.1201/9781003098256-10

TABLE 10.1   Types of Digital Legacy

Type of digital legacy	Definition	Related work
Digital remains	The bits and pieces that reflect the users' digital personality and, at the same time, compose the memories for the friends and family of the deceased	Morse and Birnhack (2020)
Digital assets	Any electronic asset of personal or economic value	Harbinja (2017a, 2017b)Rycroft (2020)
Digital traces	Digital footprints left behind through digital media	Burden and Savin-Baden (2019)
Digital endurance	The creation of a lasting digital legacy, being posthumously present through digital reanimation	Bassett (2020)

Morse and Birnack (2020) undertook a national survey of the Israeli population and found a broad range of perceptions and practices about access to digital remains. What was interesting was that their findings indicated low awareness of the available tools to manage digital remains, as well as reluctance to use them. What they noted is that platforms motivated by their own commercial interests tend to make their own policies, since there is little legislative or judicial guidance. A further difficultly with digital remains is that they continue to grow and proliferate as the number of dead people on Facebook continues to grow. Social media sites have become, as Bassett (2020: 78) points out, 'accidental' digital memory platforms and this continual growth is likely to become increasingly problematic, despite the introduction of Facebook's legacy contact appointed to manage Facebook digital estates post-death. This is partly because of the increasing number of the dead on Facebook, as well as the controversies about what should be done with any digital estate and who decides this.

Digital Assets

Digital assets relate to more tangible digital property than digital remains and tend in the main to refer to music, photographs and other assets stored on computers and clouds. There has been diverse media coverage relating to digital assets, one in particular about whether the actor Bruce Willis can leave his music collection stored on his Apple devices to his children or whether instead the collection would revert to Apple on his death. Rycroft explains:

> Digital photographs which have not been printed off, exist in a strange legal space where rights and responsibilities are often

poorly understood. Music Libraries held online have often been purchased in some way by the user but are certainly not 'owned' by the purchaser in the same way a vinyl record or CD collection would be. The purchaser of an online music library has in effect purchased a licence to download and listen to the music and does not have the same right to pass on the music to a third party that the owner of a record or CD would have.

(RYCROFT, 2020:130)

Digital assets, whilst complex to administrate, do seem to be more manageable than remains which tend to exist on social network sites that have more complex terms and conditions, compared with property stored on the deceased person's computer. Rycroft discusses a range of practical issues regarding access to the digital assets of a deceased person including that if no paper trail exists, then some digital assets may never be identified and accessed. Thus, any financial, sentimental or other such value to the loved ones will be lost. The investment that companies have put into lawyers drawing up terms and conditions in their own interests are apparently met with an uninformed 'tick here' in agreement by the Users, who are potentially unaware of what they are signing up to and whose interests they serve.

Digital Traces

Digital traces are the traces, or digital footprints left behind by interaction with digital media. These tend to be of two types: intentional digital traces – emails, texts and blog posts; and unintentional digital traces – records of website searches and logs of movements. Accidental residents of digital afterlife who leave unintentional traces are seen as internet ghosts or the 'restless dead' (Nansen et al, 2015). The somewhat eerie consequences and the impact on recipients are unclear, particularly in relation to ancestor veneration avatars, where people are immortalized as avatars in online role-playing games (Bainbridge, 2013). The traces may be intentional creations predeath, or unintentional for the dead but intentional by those left behind.

Digital Endurance

Bassett (2020) argues that 'digital endurance' is a term that captures enduring digital memories and messages posthumously and suggests that for many bereaved people this digital data contains the essence of the dead. Whilst her term refers directly to describe people's experiences of creating and inheriting digital memories and messages, it is also increasingly used in the context of creating post-death avatars so that the dead can be reanimated.

A recent study examined issues of digital legacy and legacy conditions. Cook et al. (2019) undertook a study of 32 ($n = 32$) Australians over the age of 65 and identified critical issues in the transfer, ownership, management and mobility of digital objects under legacy conditions. The authors suggest digital legacy management is a burden and that there is a need for regulatory reform in the process of making the transition from digital objects to digital assets. Studies such as this indicate there is still much legal work to be undertaken in this area of digital legacy worldwide, as well as in the areas of post-mortem ownership and post-mortem privacy.

OWNERSHIP AND PRIVACY

There is often tension between the rights of the user to privacy and the interests of the family and friends who may wish to access the digital legacy of the deceased. However, there is also an array of different terms relating to this, which are summarized in Table 10.2.

TABLE 10.2   Post-mortem Ownership and Privacy

Term	Definition	Related work
Post-mortem privacy	The right of the deceased to control their personality rights and digital remains post-mortem, broadly, or the right to privacy and data protection post-mortem	Harbinja (2017b)
Postmortal privacy	This is the protection of the informational body – the 'body' that keeps living post-mortem through the deceased's personal data, social networks, memes and other digital assets	Harbinja (2020)
Post-mortem ownership	The ownership of assets – physical and digital – by inheritors post-death	Rycroft (2020)
Digital estate planning	Organizing digital assets to ensure that these are handed over correctly to the dependents or loved ones	Rycroft (2020) Sofka (2020a)
Death-tech sites	The growth of technology connected with death and bereavement	Nansen et al. (2017)

DIGITAL ESTATE PLANNING AND
THE DEATH-TECH INDUSTRY

Many people believe that leaving a list of passwords behind is the best way forward so executors can gain access to information and assets stored on them. It is important to note that the use of someone else's password is contrary to the Computer Misuse Act 1990 Section 1, which expressly prohibits using the known password of a third party. Thus, it is important that none of this information is contained within the will since it becomes a public document. The executors should know passwords exist but not be provided with them as doing so may be breaching the terms and conditions of social media sites. The Data Protection Act (2018) relates to living persons only and terms and conditions are complex. Rycroft offers a list of useful steps to ensure that your beneficiaries can access your digital assets:

- Back up everything that is held on digital devices. If you back up into a cloud-based system think about what would happen if that failed or if you died and your subscription ended because your bank account was frozen or closed.

- Print off really important photographs and save videos offline. Technology is changing all the time and the devices we have today may not be compatible with those we have tomorrow.

- Transfer digital music libraries onto a portable device. The portable device is a physical object which can then be gifted to a beneficiary.

- Be mindful that some intellectual property rights enjoyed in life such as trademarks, patents and domain names are usually subject to renewal upon the payment of certain fees, so that these assets should be actively maintained in order to be preserved. The same may apply with regard to platforms used to store digital images such as cloud-based lockers and apps.

- Complete and keep updated 'Digital Directory', which is a list on paper of your digital assets. Clearly, such a document will help them be found upon loss of mental capacity or death and in the meantime the process of making and updating it may also help identify issues to be addressed.

- Think about any written material created online and make a decision about what is to happen post-death, who will be in charge of it and who will benefit from its value or whether you wish it all to be destroyed.

- Appoint a Digital Social Media Executor if you want a different person to have responsibility for that aspect of your digital life. Similarly utilize and engage with any protocols that digital platforms you use have available for this purpose, for example, the Facebook Legacy Contact.

- Make a will with a suitable expert and experienced solicitor so that you can include all of your wishes with regard to your digital assets in a way that is clear, unambiguous and legally binding. Note under UK law a person has to be an adult (aged 18 or over) to make a will.

(slightly adapted from Rycroft, 2020: 141)

Digital estate planning is the process of organizing your digital assets effectively to ensure that these are handed over correctly to your beneficiaries. Initially, this involves creating an inventory of digital assets and considering what dependents may need on death as well as considering what should be bestowed to whom. Digital assets include logins for social media accounts, blogs, investments, images, videos, files, messages, software, credit card reward points, bank accounts, shares, email accounts, websites and cloud storage. It is important to remember that not all digital assets can be shared, such as professional assets. Once all digital assets have been identified, they need to be stored securely and the executor named, and the executor needs to know how to locate the assets. One useful option is Cloud Locker (https://www.cloudlocker.eu/en/index.html) which is a cloud-based facility that you own and control physically as well as digitally, with patent-pending sharing and media control features. Other options include AfterVault (https://aftervault.com/) and Safebeyond (https://www.safebeyond.com/).

Death-tech Sites

The death-tech industry is the growth of technology connected with death and bereavement. Nansen et al. (2017) argue that the funeral industry is 'adopting and adapting the technological affordances, cultural norms, and affective registers of social media to remediate "traditional sites", rituals, and relations of grieving' (pp73-74). Services offered by the funeral industry include 3D printing to produce commemorative urns, QR codes for headstones and cemetery management systems that provide virtual reality tours.

Whilst death-tech sites include the possibility for design and organizing a funeral online, the most recent growth, discussed below, is a series of companies who provide a range of services from recording predeath memories to promising the possibility of an almost sentient post-death avatar of the deceased. It is noticeable that with developments in digital afterlife creation there is an increasing interest in the management of death online despite much of the current software being an embellishment of the true possibilities. For example, the use of virtual assistants, such as Siri, that provide voice and conversational interfaces, the growth of machine learning techniques to mine large data sets and the rise in the level of autonomy being given to computer-controlled systems, all represent shifts in technology that enhance the creation of digital afterlife. This section offers a summary of the developments in the death-tech industry.

MyWishes (https://www.mywishes.co.uk) is a site whose goal is to use technology to help society think about and make plans for death, illustrated in Figure 10.1. Through the provision of a care planning application, users are able to document their wishes in a number of ways. The 'online accounts' provide features to enable users to document their wishes relating to their social media sites (Facebook, Twitter, Instagram, etc.) within a digital will. This may also be linked to the person's other online accounts and state what their wishes are for each. Upon completion, the digital will is downloaded and shared with the person or persons entrusted to administer their digital estate following their death.

Empathy – http://empathy.com

This is an Israeli company, started in 2021, that helps families to manage their affairs pre- and post-death through a website and an app. The site has 150 original articles providing information and advice. The app is designed to both be supportive and provide information through a mobile phone. In practice, the company offers services for free for the first 30 days, and after that, it costs a one-off fee of $65, for as long as you wish to use the service.

Lifenaut – https://www.lifenaut.com

Lifenaut enables people to create mind files by uploading pictures, videos and documents to a digital archive. This is an explicit process that also provides a photo-based avatar of the person that will speak for them,

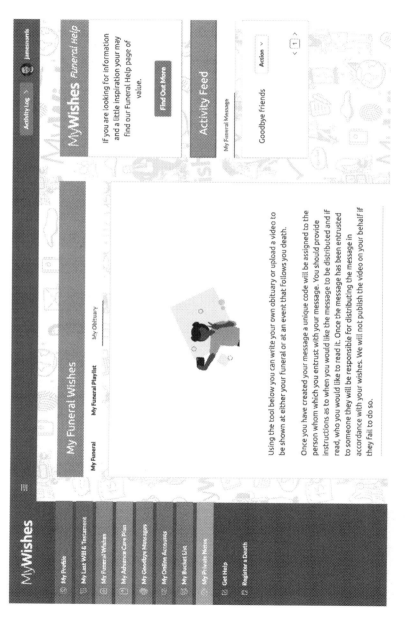

FIGURE 10.1   MyWishes website.

although currently there is the choice of only a single male and female voice which are both US English. Lifenaut is a product of the Terasem Movement Foundation (http://www.terasemmovementfoundation.com/). It is also claimed that it is possible to have your mind files beamed continually into space for later potential interception and re-creation by alien intelligences. The grandiosity of some of the claims made here is not all empirically testable. Furthermore, the idea of creating a digital archive for future use raises questions about the impact of life stage at death, as the experiences of people at different stages of the life span will impact on the legacy left behind, such that an afterlife may become 'stuck' at a particular stage without moving forward with those who are left behind.

Daden – https://www.daden.co.uk/

The work of Daden and the University of Worcester (Burden and Savin-Baden, 2019) in the UK aimed to create a virtual human, which is a digital representation of some, or all, of the looks, behaviours and interactions of a specific physical human. This prototype system contains relevant memories, knowledge, processes and personality traits of a specific real person and enables a user to interact with that virtual persona as though they were engaging with the real person. The virtual persona has the individual's subjective and possibly flawed memories, experiences and view of reality. The information provided by the individual is in a curated form and, as such, only data the individual chooses to share and be included in the virtual persona has been used so far. Whilst the project has focused on the use of such persona within the sphere of corporate knowledge management, the mere existence of the virtual persona and virtual life coaches immediately raises issues around its digital afterlife. The virtual life coach project is described in more detail in Mason-Robbie (2021).

Recordmenow App – https://recordmenow.org/

This is a free app that enables people to create a video legacy. It works by using questions to prompt people. The website offers training workshops for organizations in the areas of leaving a loving legacy, living with dying, caring for carers at end of life and supporting children, when someone close is dying. The app is easy to use and the questions used prompt reflection in different life areas, so that is possible to build a story over time with short, easy to record, videos. It can be used on a PC as well as a cell phone and also offers guidance on liturgy for different faiths. The terms and conditions do not appear to deal with issues of ownership of the digital legacy.

Aura – https://www.aura.page/

This is a site created by Paul Jameson after he was diagnosed with motor neuron disease in 2017. Paul recorded his voice font before he was no longer able to speak and in the process of dealing with this illness, he began to work with family and colleagues to develop a site that would be useful for those facing death. The focus of this site is in assisting people in creating a legacy and preparing properly to discuss, manage and celebrate life and death. The site in late 2021 comprises a series of videos to guide people, as well as useful articles on everything from funerals to virtual humans. It is early in its development, but it is striking, well presented and seems to offer a realistic stance to issues of managing the process of moving towards death and issues of digital afterlife.

Elixir – https://www.elixirforever.com/

This US company argues that before the COVID 19 most people spent 7 hours a week online and that this has grown since then. What they point out is that post-death, social media sites may freeze accounts – such as Instagram. Facebook offers a legacy contact to manage the site or the option to memorialize it and Twitter accounts can be frozen if proof of death is shown. The strapline is 'Dedicated to using AI and neural network technology to empower your digital AI self in this lifetime'. The company suggests that we need to create an authentic digital copy to enable our family and friends to continue to connect with us. The solution they propose is for us to use social media to create our own AI being which has a 'digitally positive carbon negative foot print', but the website does not say how this will be achieved.

Ankhlabs – https://ankhlabs.io/beankh/

This German company states that it brings together artificial intelligence, big data and blockchain technology to 'add immutability' to digital immortality, through an app entitled BeANKH. The BeANKH app applies AI to social media such as face recognition, emails communication, and bot and blockchain technology to build your own look-a-like digital assistant 3D avatar which uses a digital copy of your voice. They suggest it is possible to preserve and share people's essential traits and personalities long after their physical death. The use cases they cite vary from what would appear to be using a bot as a personal assistant to more innovative options. The BeANKH app enables you to undertake activities in day-to-day life, such as its Social Media Manager giving your digital copy the ability to like and comment for you on social media. There are other services which overlap with other death-tech sites, for example, you can arrange to record messages that can then be sent on a certain date in the future or after death, and you can also transfer cryptocurrencies in your BeANKH Secure Wallet with a handover date post-death. Yet despite the options for creating post-death recording and money transfers, there does not appear to be any support or link to bereavement services, as with other sites.

Iteranl.life – https://iternal.life/

This site focuses on using storytelling to create memories, very much like Recordmenow. Having signed up and provided personal details you are able to record your memories. The focus here is on helping families save their memories for generations to come, so it is really a site where you can record and save memories and probably little more than this.

Heaven Address – https://www.heavenaddress.com/

This site is linked with a mobile app entitled Memories, on which you can both record memories in order to create a digital legacy as well as providing a space to create a memorial to someone who has died. The app is linked to the Heaven Address website and the site aggregates all memorial pages and these are co-branded and linked with the relevant memorial funeral director. Nansen et al. (2017) note that this site hosts all memorial pages within the Heaven Address URL, with the aim of maximizing the use of Google's PageRank algorithm in order to ensure traffic to Heaven Address' website.

Digital Immortality Now – http://digital-immortality-now.com/

This site's strapline is 'The instrument to reach immortality for everyone'. The site suggests that it is possible to provide cheap and affordable digital immortality for everyone. It seems to be a site where people upload self-recording and self-description to facilitate future reconstruction. The site provides a long article on digital immortality and the possibilities for the future but does not really explain how this will be undertaken. Some of the steps suggested to create immortality include collecting your DNA, collecting existing data such as photos, documents and video, extracting personal data from social networks, running psychological tests, interviewing friends and creating art. The steps suggested seem little different from other sites, for example, they advocate a number of steps to achieve digital immortality creation – which in fact is really just the creation of a digital legacy. Other parts of the site seem to be presenting futuristic transhumanism, with articles to support this stance. It is clear that current death-tech sites range from basic sites that just offer streamed funeral services with memorial pages to sites that provide a complete planning package with tutorials and social media support. However, few provide digital afterlife creation options; the focus tends to be on planning and memorialization.

Since there is an increase in streaming funeral services during 2020–21, particular new software has been developed to broadcast funerals. Some of these services such as OneRoom (https://www.oneroomstreaming.com/) in Australia will only provide the service via funeral directors. In the UK, Funeral streaming (https://www.funeralstreaming.co.uk/) provides choices that include streaming it yourself for £95 or a professional package for just under £600. It has been noted by Kuipers (2021) that in the US some families are choosing a streamed funeral over a live event since it is cheaper.

ETHICAL RESPONSIBILITY

Ethical concerns stretch beyond the legal concerns since digital memorialization relates to the social good and thus introduces questions about what counts as the social good and who decides? Can or should everyone have the right to a post-death avatar? What will occur if such avatars become sentient? Are there different levels of digital memorialization that require different levels of ethical stances? Currently, there is not a single global jurisdiction, and the laws that apply in one country or states may not necessarily apply elsewhere, which means that the use and abuse of virtual humans across the world, particularly in terms of those who create virtual humans for illegal purposes for use in a different country from their own, are unlikely to be prosecuted. Furthermore, it is not clear if the actions of the virtual human may result in criminal liability. This raises the question again about the driving force behind the virtual presence; the dead can generally have no criminal liability, but others acting behind the shield of a digital presence may have. Questions need to be considered such as

- What if autonomous military combat robots or other AI-enabled technology destroy or kills unintended targets?

- What if a claimant were to attempt an action in libel against a virtual presence?

- If a virtual human was acting under the deceased's autonomous control, for example, a perpetuating social media presence, can a claimant attempt to sue the dead persons' estate?

Digital immortals are designed using human decision criteria and therefore ethical behaviours need to be 'designed-in' to them. Riek and Howard (2014) suggest that design considerations should include:

- reasonable transparency in the programming of the systems

- predictability in their behaviour

- trustworthy system design principles across hardware and software design and

- opt-out mechanisms (kill switches)

It is important to consider what occurs if AI-enabled technology mistakenly attacks incorrect targets or fails to distinguish correct targets or in the case of a self-driving car, makes an incorrect decision. Further, in the context of the range of virtual humans that have been developed already, particularly those being designed for use pre and post-death, designing appropriate and effective ethical standards remains complex and far-reaching

Facebook uses machine learning to detect the possible intent of suicide (Gomes de Andrade et al, 2018). The features they focus on are the content of the posts, the time of day and reactions to the posts; these classify the range of likelihood of suicidal intent. Gomes de Andrade et al. possess expertise in the area of suicide prevention and considered ethical concerns such as:

- whether Facebook should be developing and deploying prevention tools

- balancing the issue of privacy and efficacy, they argue:

> When building suicide prevention tools, one of the balances we need to attain is between efficacy and privacy. These interests may be at odds with each other, as going too far in efficacy—detecting suicidal ideation at all costs, with no regards to limits or boundaries—could compromise or undermine privacy,

i.e., the control over your own information and how it is used. The question we were faced with was the following: How can we deploy a suicide prevention system that is effective, and that protects people's privacy, i.e., that is not intrusive and respectful of people's privacy expectations?

(GOMES DE ANDRADE ET AL., 2018: 679)

- the use of machine learning intervention versus human intervention

- the choice of using machine learning thresholds to decide on the level of intervention

The Facebook intervention offers an interesting and useful illustration of how artificial intelligence can be used for suicide prevention, but in practice, artificial intelligence intervention does not deal well with the ethical complexities involved. However, the use of artificial intelligence to deal with such a complex human problem is not a straightforward solution, although it could be a help to some degree.

The Location of the Dead

Nansen et al. (2017) argue that by accessing a memorial on a mobile app the deceased is removed from a physical place of sequestration and is then located in an imported social network. Whilst this stance might reflect the behaviours of those who speak to the dead in spaces like Facebook as if they were there, this does introduce questions about where people believe the dead to be situated. The perception of the location of the dead relates largely to spiritual beliefs about the afterlife, whether the dead have sentience, whether people believe in the notion of souls and even if the dead can be harmed posthumously. Stokes (2015) questions the extent to which digital artefacts have an impact on the ontological and ethical status of the dead. He argues that:

the dead persist as objects of moral obligation precisely because in memory we give them the same phenomenal presence that made them morally compelling while they lived.

(STOKES, 2015: 238)

One of the interesting concerns Stokes raises is whether the dead can be harmed by events after their death. Whilst this can be considered from a number of philosophical positions, from a spiritual stance it would appear that the dead could be harmed. For example, in Buddhism, if the dead are not prayed for, it may affect reincarnation, or in Hinduism, if their social media profile is trolled, this may affect their status in the afterlife. While Stokes suggests the need for a distinction between self and the person, what he really seems to be suggesting is a distinction between a body and a soul. Furthermore, he suggests that whilst new technologies may make it easier to remember the dead, at the same time we may 'lose precisely the dimension of lostness that should characterize our relationship with the dead' (p 246). This in turn introduces ethical questions about left-behind identities and information.

Left-Behind Identities and Information

The issue of what is left behind and who controls the deceased's digital assets remains a contested area. In law, the right to delete social media profiles and information is still unclear and few (if any) provide conflict resolution for family disputes in relation to memorial sites. Wright has noted that 'protecting the privacy of deceased users is not a legally valid concern' (Wright, 2014). Thus it remains questionable whether the dead have a right to privacy but it is clear that if the dead person has not agreed in advance to have their profile memorialized, this in itself raises ethical concerns. For example, family members who may not have been the deceased's friends on Facebook may be unable to access any memorials created on that site. Furthermore, if they are unfamiliar with social media practices they may be upset by the comments made on such sites.

A recent innovation is the creation of the Web 2.0 Suicide Machine (http://suicidemachine.org/), which offers automated deletion of contacts and content on social media sites and is a site that offers what Lagerkvist (2017) terms 'media end'. Arguing that everyone should have the option to commit digital suicide, this site's states:

This machine lets you delete all your energy sucking social-networking profiles, kill your fake virtual friends, and completely do away with your Web2.0 alterego. The machine is just a metaphor for the website which moddr_ is hosting; the belly of the beast where the web2.0 suicide scripts are maintained. Our service currently runs with Facebook, Myspace, Twitter and LinkedIn! Commit NOW!

There is little research on the impact of this to date or the implication for this in relation to death and suicide in real life. The use of the term 'suicide' in the context of deletion of profiles is both insensitive and unhelpful. However, the existence of the site does illustrate how people become addicted and entangled in social media sites and networks to the extent that they feel the need to leave completely. Interestingly Facebook has blocked the Suicide Machine from their servers because they argue that it contravenes their Statement of Rights and Responsibilities.

Family Decisions and Conflict

It is clear that there is potential tension between the rights of the user to privacy and the interests of the family and friends who may wish to access the digital assets of their loved ones for a variety of reasons. Access rights to online material after a person has died remains a challenge when passwords are not known. The Data Protection Act (1998) and the GDPR (2018) relate to living persons only, and as terms and conditions are overly complex and too many, unintelligible, there are also concerns about the extent to which relatives are able to understand and access the social media sites that were being used, since there are still people who lack digital fluency. For example, there is particular online behaviour and etiquette that is expected on memorial sites and since the use of language changes over different generations, terms that may be used by one generation might be misinterpreted by those of another.

Whilst the notion of digital fluency tends largely to be centred on people and their capabilities, at the same time there has been an increasing fluidity in the realms of technology. The emergence of social network sites in the early 2000s promoted a high degree of interest and interaction but sites in the main remained separate. Now, in 2022, sites are more rhizomatic so that users share their profiles across sites as well as collating their data from apps there, whether it be sports information or diet trackers. This brings with it challenges about data sharing. For example, people's contacts become available across a wide range of sites and in 2007 the 'social' graph began to be recognized as a key marketing tool. The social graph is defined as the links between people in a system (Fitzpatrick and Recordon, 2007) and is now seen as highly valuable for advertising, since third-party developers can build software on top of the social graph, so connecting people to advertising and other sites via their Facebook Friends list. Digital fluency then has been aided by changes in technology, making it easier to communicate and share information with others. At the same time, engagement with social networking sites has become more media-centric than in former years. Today activities, photographs and achievements are shared, rather than profiles. Such interaction with technologies in these kinds of ways introduces questions about the power relationship between the users and the corporate producers, particularly in the death-tech industry. Digital fluency at one level is now no longer about mere net savviness but rather can be seen as a deeply contested marketing tool, in terms of the ways organizations configure their technology for their users and the impact this has on the moral status of media post-death.

CONCLUSION

Digital legacy remains a complex and contested area with inadequate legal guidance and jurisdiction. There is also little in-depth understanding by many people about the challenges of post-mortem privacy, ownership and the rights of different family members. Furthermore, the moral impact of post-mortem harm raises interesting dilemmas for those with a faith. At the same time, the confusion about where the dead are located remains an area that is troublesome, not only to the general public but also to experts in death studies and members of the religious community, as was seen in Chapter 9. Despite these legal, moral and spiritual conundrums, death in other contexts has become spectacular, as will be seen next in Chapter 11.

Ambivalence and Spectacle

INTRODUCTION

This chapter begins by exploring the idea of spectacular death and suggesting that the use of artificial intelligence has prompted the development of spectacular death as well as ambivalence towards it. The chapter then examines the commercialization and exploitation of death suggesting that this has resulted in the need for crowdfunded funerals. The next section of the chapter analyses the idea of the ambivalent relationship people have towards their digital devices. The final part of the chapter explores the growth and change, and the impact of dark tourism, including the rise of immersive dark tourism and the changing landscape of thanatopathia and thanatopolitics.

AMBIVALENCE AND SPECTACLE

It is clear that there is a huge ambivalence towards the use of artificial intelligence in the management of dying and death. Such ambivalence is seen in the portrayal of young superheroes' deaths in film, the use of dead celebrities in advertising and films and the use of algorithmically based apps to predict death. A further sense of ambivalence is seen in ways in which death in the 2020s has become increasingly spectacular. Spectacular death is defined by Jacobson (2021) as the new attitude to death in contemporary society, whereby death has been transformed into a spectacle. What is particularly helpful about this work by Jacobsen is the

DOI: 10.1201/9781003098256-11

way in which he suggests that there are a number of paradoxes as death lingers between autonomy and control, and liberation and denial, which I suggest reflects the contemporary ambivalence towards death in a digital age. Further, that idea of death as spectacular also reflects the shift in the 21st-century society away from death and spirituality towards the importance of the role of media in providing information and moral orientation towards death (Hjarvard, 2008). Much of this seems to relate to the way in which, as Jacobsen (2021) argues, that death has become spectacular. He suggests that this occurs in five ways, as summarized below:

1. *The mediation/mediatization of death.* This is the way that both real death and fictional death are central to our media landscape through television, the internet, films, literature and popular culture.

2. *The commercialization/ commodification of death.* Death has become a standardized event in our society and therefore is necessarily commodified by those who provide ceremony experiences, those who manage the body, as well as therapists and counsellors.

3. *The de-ritualization of death* is exemplified in new forms of rituals and expression. This is also seen in the blurring of the boundary between commercialization and ritualization in consumer capitalism.

4. *The palliative care revolution* has reformed and refocused the way death dying and bereavement is managed so that death is recognized as is the importance of dignity in death.

5. *The specialization of death* emerged from the palliative care revolution but now extends into academic areas such as thanatology, death studies and the sociology of death. The death awareness movement has shifted death away from being a taboo topic.

Although Jacobsen does not refer directly to artificial intelligence in these five delineations some of these forms of spectacular death do relate directly to digital afterlife such as the mediation/mediatization of death, the commercialization of death and the de-ritualization of death in digital spaces.

COMMERCIALIZATION AND CROWDFUNDING

The commercialization of death has been growing over the last 50 years, with the result that there have been increasing concerns about the ethos of the funeral industry. Jacobsen (2021) notes that the paid work of the funeral industry in the 20th century has ushered in new forms of funeral rites and opportunities for making money. He argues that we have moved towards a need for a much more personalized funeral event. Thus, we are seeing a Disneyized way of creating enchantment and emotional attachment in death and mourning rituals rather than the McDonaldized standardized pre-packaged options of former years (p.11).

A further form of commercialization is seen in the growth of the digital legacy industry as was discussed in Chapter 10. Digital legacy companies offer a range of options for post-death preservation, digital legacy management and digital immortality. These companies are not yet big commercial players as this industry has not really taken off. However, as Kasket (2021: 25) suggests, social media are increasingly becoming the new undertakers of people's digital remains. Kasket cites Facebook as just such an example of this especially since both Facebook and Instagram do not show advertising related to memorialized profiles. However, it is evident according to Öhman and Watson (2019) that social media with deceased use status can be mined for artificial intelligence modelling and to gain insight into the market. They argue that memorialized profiles may attract living users to the relevant site and that datasets of digital remains may provide historical insights, which could offer a valuable market advantage. Thus, what we are seeing is the increasing growth of the use of machine learning in covert ways in the context of people's loss and grief. At the same time, the funeral business has become a profitable industry, which in recent years has garnered the use of technology to enhance its provision and ensure the marketization of death.

The commercialization of the funeral event itself has risen exponentially in 2020 due to the COVID-19 pandemic. Yet, even before this, the cost of funerals in the UK ranged between £3,000 and £5,000 for a funeral, rising by more than two-thirds since 2010 (Kollowe, 2018). Whilst it is possible to save over £1,000 by comparing funeral services in their area, people tended not to do this, and since funeral costs are invariably not online, this makes researching in times of grief more complicated. The result of the COVID-19 pandemic has been the live streaming and recording of what is often a service of between 15 and 30 minutes long. While some churches provide this as a free service, most charge around £50–80 and upload the service with a private link to Facebook. Other webcasting businesses are increasing their revenue through specific costing by providing options as follows:

- £30 to watch the service

- £45 for downloadable link lasting 28 days

- £50 for physical copy

However, many websites do not provide a clear price list and hidden costs can increase expenses dramatically causing decision-making process to be difficult in times of grief. The commercialization of the death industry has also resulted in the need for crowdfunded funerals. Kneese (2018) undertook a study to examine crowdfunded funeral campaigns. These campaigns illustrate the many inequalities which are often hidden in digital death and mourning practices. Kneese (2018: 1) cites the example of the shooting of Michael Brown, an unarmed African-American teenager by a white police officer. A clerk working for the Brown family's lawyers initiated a GoFundMe crowdfunding campaign. Although the funds fell short of the amount needed, the 42,000 individuals shared the crowdfunding page via Facebook and Twitter. This incident and the subsequent crowdfunded funeral raised national awareness of movements like Black Lives Matter. The founders of Black Lives Matter, three Black women, Alicia Garza, Patrisse Cullors and Opal Tometi have always put LGBTQ voices at the centre of the conversation. Since the crowdfunded funeral of Michael Brown, further ones have been initiated in the wake of other police shootings, resulting in the Black Lives Matter movement making the political potential of crowdfunded funerals especially visible.

Crowfunded funeral campaigns use stories and images through social networks, and in order to garner funds, many of these have political undertones. Although this method of funding funerals is vital and important, this has a commercial side to it as well as Kneese explains:

> some ethical questions are raised by crowdfunded funerals, where corporate platforms owners profit from heartache and tragedy. For instance, GoFundMe takes 5% of every donation it receives (Han, 2015). Money for funerals and memorial services is raised on general crowdfunding platforms like DonationTo, IndieGogo, YouCaring, and GoFundMe, as well as funeral-specific companies like FuneralFund and GracefulGoodbye. It is difficult to verify where money raised by such campaigns goes, since sites like GoFundMe do not investigate cases of potential fraud (Han, 2015).

(KNEESE, 2018: 2)

Whilst Kneese's study was based in the US, there has been a rise in the use of crowdfunded funerals in the UK using sites such as JustGiving and Beyond, which currently do not appear to charge. Apart from commercialization within the death industry, other forms of ambivalence can be seen in the deployment of objects and devices, not originally designed to support grief and mourning.

TRANSITIONAL OBJECTS AND AMBIVALENT DEVICES

This section examines the idea that transitional objects, in particular mobile phones, raise interesting conundrums about the relationship between death and the digital. The ambivalence towards phones is because they are seen as a combination of ritual companions, portable cemeteries and prostheses – extensions of ourselves.

Drawing on the work of Winnicott (1953, 1960), Mowlabocus (2016) explains that a transitional object is one that provides a bridge between the subjective self of the child and the external world. Such an object, a favourite toy or blanket, provides not only a realization of the self but also comfort and reassurance, particularly in times of unfamiliarity. Earlier work on phones as transitional objects focused on them being devices that facilitated the mediation of relationships (Keefer et al, 2012), yet recent work has also introduced the issues that loss or detachment from mobile phones can result in NOMOPHOBIA or NO MObile PHone PhoBIA, a psychological condition when people have a fear of being detached from mobile phone connectivity. However, Mowlabocus suggests that it is important to consider the relationship we form with the phone. She suggests that we turn to our phones in-between spaces or activities, often when we are uncertain about how to occupy our time or be in a particular space. The sense of being in an in-between space has similarities with a sense of uncertainty which occurs after the death of a loved one. The phones of the deceased appear to become a source of comfort, something, or 'someone' to hold, as well as a form of distraction from the actuality of death. Yet, I suggest that phones are also seen as ambivalent devices in the face of death. du Preez (2018: 2) argues for the notion of sublime selfies 'Selfies unknowingly taken before death, selfies of death where the taker's death is almost witnessed, and selfies with death where the taker stands by while someone else dies'.

However, what is perhaps more of a concern is the notion of selfiecide – dying while taking a selfie, an example of which was a selfie by the young

Russian girl Xenia Ignatyeva in April 2014, who climbed on to a high bridge to impress her friends but then slipped and fell and was killed by electric fences. Since 2013, at least 259 people have met an untimely death while attempting to take a selfie, most of them being young men taking selfies from precarious cliffs or buildings (The Star, 2019). What we appear to be seeing in the practice of selfiecide is the celebration of risk in a time of chronic uncertainly. This could be seen as *anomie* – Durkheim's term to define the breakdown of morals, values and standards, particularly since his first use of the terms referred to 'derangement', a sense of desire without limits (Durkheim, 1893). Beck might argue that chronic uncertainty is reflected through a risk society. Reflexive modernization, the process by which the classical industrial society has modernized itself, has resulted in a sense of crisis characterized by a 'risk society' (Beck, 1992). This type of society, with its emerging themes of ecological safety, the danger of losing control over scientific and technological innovations and the growth of a more flexible labour force, has and continues to have a profound effect on society and, more recently, death in society. This desire to take risks and survive are resulting in societal ambivalence towards the value of the epic selfie at the cost of death. Indeed, as Bauman argued in the 1980s, we have become a risk society – and that includes risking our death at the expense of our own self-media coverage (Beck, 1992). Thus what I suggest we are seeing is a number of different forms of selfies, which I term 'ambivalent selfies', as illustrated in Table 11.1.

TABLE 11.1   Ambivalent Selfies

Type of selfie	Definition	Related work
Before death	Taken before imminent death, such as an aeroplane crash	du Preez (2018)
Of death	Usually accidental, such as falling off a bridge by mistake	du Preez (2018)
With death	Taken with a dead person at a funeral home or at a funeral, such as a picture of oneself with the coffin in the background	du Preez (2018)
As tragedy	Taken with a tragedy in the background, either as the watcher or, in the case of the Sewol disaster, as one's own drowning tragedy	Hjorth and Cumisky (2018)
As witness	As a witness to a live event such as a shooting or a suicide	Hjorth and Cumisky (2018)
As eulogy	The phone was used as a eulogy of/for the deceased containing footage of selfies before they died	Hjorth and Cumisky (2019)

Further ambivalence is seen in the use of phones as transitional objects since they interrupt rituals and accepted practices as they are used to share, spread and manipulate the context of death events. Papailias (2016) uses the term 'mediated witnessing' to reflect the way in which the circulation of images of traumatic events result in powerful effective genres that are somewhat haunting and fail to leave virtual witnesses untouched. This is because although the physical body is invisible, meanings, mannerisms, behaviours and unstated assumptions are clearly visible in the left behind's communication. Thus mediated witnessing results in 'context collapse' as coined by boyd, who has argued:

> that the contextual information that they draw from does not have the same implications online. Situational context can be collapsed with ease, thereby exposing an individual in an out-of-context manner.
>
> (BOYD, 2002: 12)

Whilst contexts may have collapsed, as boyd suggests in 2002, some years later it is evident that those using social networking sites do increasingly have a sense of their audience. Yet at the same time, the varied, mainstream and often linked social media sites are places and spaces that could be said to be resulting in multiple, though perhaps different forms of, context collapse. Mourning practices are becoming new types of both performance and cartography on tour. Lammes suggests in gaming, players explore and master environments through digital mapping. The player becomes an imaginary cartographer while creating a spatial story around himself/herself so that maps and tours necessitate one another and come into being through two-way movement (Lammes, 2008: 87–88). Thus mediated mourning practices through the use of phones and social networking sites result in context collapse between the sites, but also result in mourning being a tour around such spaces, thereby creating virtual maps for the mourners.

Thus, when witnessing collapses the mourner witnesses a divide as mourning and witnessing become entangled. For example, Hjorth and Cumiskey (2019) narrate the story of the Sewol disaster in Korea and the way schoolgirl Park Ye-seul and her friends filmed the disaster as selfie videos, which epitomize mediated witnessing as they explain:

> The video conversation, which can be found on YouTube, consists of a conversation between Ye-seul and her fellow passengers as

well as her copresent parents. She talks of how scared the other passengers are while begging, 'Please rescue us'. They talk about the increasing tilt of the boat. Then there is an official announcement: 'Please double check your life jacket whether it was tighten well or not. Please check and tighten it again'. Ye- seul says to her videoing phone (as if her parents are inside it), 'Oh we're going to diving into the water', followed by 'Mum, I am so sorry. Sorry Dad! We will be okay! See you alive'.

Her father recovered the camera phone footage after her death. He dried it out and replaced the SIM card. In the phone were videos she had filmed before she died. For her father, his daughter's phone was not just a vessel for channelling a re-enactment of his daughter's last moments alive, but also in doing so it afforded him the ability to move back in time and space to be 'present' with her during her last moments.

(HJORTH AND CUMISKEY, 2019: 191)

While Mowlabocus (2016) suggests the phone as being a transitional object in relation to comfort, it is also an object that might be seen as being transitional in other ways. For example, in the transitional phases of grieving by maintaining continuing bonds or in a mediating relationship with the dead. Cumiskey and Hjorth (2018) argue that mobile phone use in the context of mourning is highly ritualized and personalized. The result is that phones become companions and part of the ritual of grieving. This is also seen in the study undertaken by O'Connor in 2020, who narrates the story of a research participant whose sister died and who had changed her phone shortly before her sister's death. The conundrum for the left behind participants was whether to turn on the old mobile phone and look through the texts. O'Connor suggests that the possibility of using digital material to create an afterlife for the deceased is not something that is chosen by those left behind. Instead, the digital is blended with history and memory, and the mobile phone is left turned off in a drawer.

Apart from the need to hold the phone and send messages to the dead, mobiles are also often seen as extensions of the self and indeed extensions of the deceased. Hjorth and Cumiskey (2018: 175) argue the 'mobile device itself becomes an extension of the self and a mechanism through which it can at once service a lifeline and act as a portal for the extension of one's presence and influence beyond death'. In terms of digital afterlife and the spiritual realm the selfie, like many other forms of digital legacy outlives the time and space in which it was created. Senft and Baym (2015: 1589) see the selfie as both an object and a gesture. The object 'initiates the transmission of human feeling' and as a gesture, it sends different messages to different individuals. The mobile is also used as a repository for the deceased; it becomes not only an extension of the dead but also a eulogy. Yet the use of the phone for taking selfies has now become part of dark tourism and this brings with it a different kind of ambivalence.

DARK TOURISM

Dark tourism has been something that has been in evidence for many years, whether the spectacle of hanging or witch burning. More recent examples of this are tours around famous murder sites and the London death dungeons. Dark tourism in the main still larger relates to 'the presentation and consumption (by visitors) of real and commodified death and disaster sites' (Foley and Lennon, 1996:198). Since 1998, there have been ensuring arguments for different forms of dark tourism. These range from the idea of dark and darker tourism (Miles, 2002) and suggesting a distinction between sites associated with death and sites of death. Thus, for example, visiting Auschwitz/Birkenau is darker since it is a site of death, compared with visiting a Holocaust Memorial museum, which is associated with or is memorializing death. Further, Stone (2006) has suggested that there are shades of dark tourism, and based on his typology and adapted from his work, the following examples have been created.

- Type 1 Lightest: The focus is on the forms of sanitized entertainment designed to scare, such as the London Dungeon, the Dark Fun Factory and the Tower of London.

- Type 2 Lighter: Dark Exhibitions that are educational and often commemorative, such as the Twin Towers exhibition at the Smithsonian. However, some are more primitive and to some degree commercial, such as the 'Body Worlds' anatomical displays of human bodies that have been preserved. Lighter types can also include battle re-enactments at sites such as the Bosworth battlefield site in Warwickshire, UK.

- Type 3 Light: Dark Dungeons are sites that are based on former prisons and courtrooms. An example would be Bodmin Jail in the UK, or the Galleries of Justice in Nottingham, UK, which markets itself as the only place where you could be arrested, sentenced and executed.

- To some extent, Robben Island in South Africa could be said to fit into this category (according to Stone, 2006) but I would argue that the political dimension of this site probably fits more appropriately into the Dark category.

- Type 4 Dark: Dark Resting Places and Dark Shrines are where graves or cemeteries are seen as spaces of dark tourism. An example of this would be Highgate cemetery in London, which is both romanticized and often used on film sets, which results in it being a tourist attraction. However, as Stone (2006: 155) explains they also 'revolve around a history-centric, conservational and commemorative ethic'. In contrast, Dark Shrines are those sites that focus on remembering the recently deceased; examples of this are roadside shrines and they could also be media shrines such as Facebook shrines created by fans for recently deceased celebrities. The latter perhaps also overlap with Type 5, Darker.

- Type 5 Darker: Dark Conflict Sites such as battlefields and ex-war zones with significant bloodshed or associated horrors that have both an educational and commemorative focus. These range from First World War sites in France to sites such as the Soviet forced labour camps of the Gulag, which also overlap with the darkest category.

- Type 6 Darkest: Dark Camps of Genocide are sites where genocide and atrocities have occurred in places such as Rwanda, Cambodia and Kosovo and holocaust sites such as Auschwitz/Birkenau. As Stone (2006: 157) notes, holocaust-based sites and exhibitions often dominate general dark tourism discussions and are positioned at the 'darkest' edge of the 'dark tourism spectrum', as illustrated in Figure 11.1.

This typology is helpful in understanding different and diverse types of dark tourism and provides a backdrop for new forms of dark tourism in the context of immersive forms, which will be discussed next.

More recent work in this area of dark tourism discusses the recreation of macabre selfies. Hodalska (2019) suggests that, whilst photography of places of death has occurred for many years, there has been a change in recent years in the photos taken so that whilst media, in the main, report tragedy, dark tourism tends to result in the commodification of death sites. What is particularly poignant about the practice of taking selfies at sites of death is the ambivalence and the confusion associated with

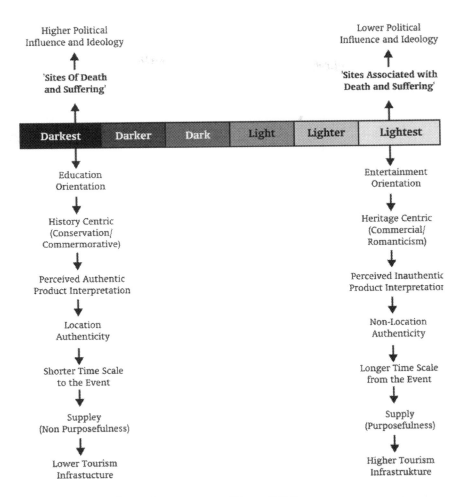

FIGURE 11.1   Dark tourism spectrum (Stone, 2006).

this behaviour. Hodalska (2019: 413) documents how teenagers posed for selfies at mass murder sites such as Auschwitz and posted them on Facebook, suggesting they were there to have fun. However, she asked:

> Maybe the youngsters are demonstrating rebellion against mandatory school trips to memorial sites, they act out and use technology as a platform to express their feelings? Maybe in places of death teenagers feel the urge to celebrate life in, what seems to them, the most 'creative' way: posing by the railway tracks leading to the camp, with their thumbs raised, almost asking for a fun ride.

Another development is that of immersive dark tourism. Wright (2021) argues that Western attitudes will shift towards the age of immersive death. What Wright suggests is that new forms of immersive technology will enable people to explore their own death, from the darkest form to lighter forms provided in death entertainment. Wright suggests three virtual reality (VR) dark attractions:

- The inside of Auschwitz VR which is guided by stories and supported by a 360-degree drone; users can experience a 3D experience

- The Anne Frank house VR which offers users an emotional insight into the two years Anne Frank spent in this house during the Second World War

- The Berlin Wall VR where you can go behind the wall to experience what communist East Berlin was like at the height of the Cold War

What is interesting about Wright's work is that the encounters with virtual reality in death are those of historical door dark tourism, rather than the possibilities for an exploration of one's own death. Whilst these are clearly immersive experiences, what they do not provide is an immersive exploration of death. Barberia et al. (2018) undertook a study that examined peoples' responses to death in an immersive environment. The experiment had two groups, one who experienced a VR island and one who did not. Those chosen for the immersive experience were embodied in alternate bodies with two companions. Over time, each participant witnessed the death of the two companions and then her own death, followed by a period of observation of the continuing activities in the virtual world on an external screen. The findings suggest that those who experienced the island, report more life attitude changes, becoming more concerned with others and more interested in global rather than material issues compared to the control group. Confronting death in a more honest and realistic way such as this may be helpful, possibly more so than the historical immersive death experiences suggested by Wright (2021). Nonetheless, it is clear that the fascination with death through immersive experience also reflects the 21st-century fascination with violent death.

THANATOPATHIA AND THANATOPOLITCS

Thanatopathia is the passion for violent entertainment that is focused on death, so that death becomes a trendy spectacular commodity as well as popular entertainment. Khapaeva (2021) coined the term 'thanatopathia', from the Greek θάνατος (*thanatos*, death) and πάθος (*páthos*, desire) to reflect the ways in which the desire for violent entertainment has become central to popular culture. She suggests that this is not only evident in popular culture but also evident in the practices and rituals associated with death. Khapaeva argues that the portrayal of humans in apocalyptic and post-apocalyptic films of the 21st century is an antihuman stance. Thus, the idealized hero/heroine is the non-human and the annihilation of non-humans is seen as a natural rather than an ultimate tragedy. Such contempt for humanity, she suggests, results in the viewers' seeing the violent death of humanity as entertainment, which in turn can be linked to a reinterpretation of the Anthropocene. (The Anthropocene is defined as the earth's more recent geological time period when human activity started to have a significant impact on the planet's climate and ecosystems.) Such a reinterpretation sees the Anthropocene not as a period in history where humanity has affected and destroyed the planet along with the need for global environmental change, nor indeed as the geological period in which humans have significantly modified the planet through global warming, carbon emission and habitat destruction. Instead, it is seen as more of an antihuman philosophy, resulting in social movements such as The Voluntary Human Extinction Movement and the Dark Mountain project. The Voluntary Human Extinction Movement argues for the importance of phasing out the human race and calls for all people to abstain from reproduction to cause the gradual voluntary extinction of humankind. The Dark Mountain project brings together writers, artists and thinkers who foretell the collapse of the world in order to attempt to address, halt and reverse the destruction of the planet. Although Khapaeva's stance seems to be quite stark and negative it does introduce questions about the extent to which films do indeed change death into something both more fictional and more spectacular than former years. However, perhaps as Hamonic (2017) suggests apocalyptic and post-apocalyptic narratives can help people to cope with the fear of death, make sense of the world and encourage them to change their ways.

Thanatopolitics is the politics of death, which in many ways does reflect many of the issues raised in the newer idea of thanatopathia in that while thanatopolitics is an exploration of the politics of death in thanatopathia is also a sense of fascination with the way in which the media portray death, often in a sensationalized form. A concept originally suggested by Murray (2006) thanatopolitics is 'both a response and a resistance to biopolitical power and to the Western conception of rational sovereignty with which biopolitics is allied' (p.195). For example, Morse (2021) argues this shift of the role of the media in providing a moral orientation towards death has resulted in it becoming a moral agent through the ways in which violent death is presented and concealed (p129). Further, social media he suggests in the process of replacing traditional news at the same time depoliticizes is death. He makes a salient point about the relationship between social media and death portrayal, suggesting that there has been a changing of the guards so that journalists whose role was to contextualize death are being replaced or challenged by laypeople and algorithms. He argues:

> Given the negative consequences of irresponsible distribution of death imagery, internet platforms should consider employing journalists as facilitators of information distribution, or at least training their coders to incorporate responsible journalistic prac-tices and to teach their AI tools and algorithms to do the same.
>
> (MORSE, 2021: 139)

It is clear that some of the concerns related to ambivalence and spectacle have emerged from the lack of control of images, portals and practices related to death and the management of death in the media. Whilst Wright (2021) has suggested that immersion offers opportunities to play with and explore death, to date there is little available to explore one's own death in realistic immersive ways. Instead, what we are seeing is experiencing death through others – either through a fascination with violent death or through exploring death historically through dark tourism. It is clear that global perspectives on death have become spectacular and ambivalent but also ones where deep engagement with death is still avoided.

CONCLUSION

Most societies continue to have an ambivalent relationship with death and this has become complicated through digital media. Some of the more recent developments in the areas of dark tourism, selfiecide and thanatopathia are ones that are disturbing and thus introduce questions about acceptable norms and ethical disquiet. The use of the phone as a tool of remembrance, as a witnessing device and a source of comfort has resulted in it becoming a component of death rituals and practices. The shift towards spectacular death, and the ambivalence evident towards the digital, remains a societal challenge and ethical conundrum. However, in order to manage this ambivalence and sense of the spectacle well, it is important to set this in the context of future concerns, as will be explored in the final chapter.

CHAPTER 12

The Final Cut

INTRODUCTION

The research and literature across the topics of digital afterlife and the spiritual realm raise a diverse range of challenges. What is clear is that digital afterlife is rarely linked with notions of the spiritual realm. Indeed, for many people of faith, the digital is largely overlooked, ignored or seen as an irrelevance; and the relationship between the afterlife and spiritual realms is also often seen as being different spaces and places. This chapter begins by exploring the relationship between physical and digital death using the metaphor of the final cut. It then examines the impact of absent presence on the living and then the notions of the digital death pragmeme and finally reflects on the ongoing confusion and conundrums related to digital afterlife and the spiritual realm.

THE FINAL CUT

The Final Cut is used here with reference to Shakespeare's *Julius Caesar* (1599/2008): 'This was the most unkindest cut of all' (Act 2 scene 2), spoken by Marc Antony in his speech to the multitudes following Caesar's murder. Here Shakespeare is using ambiguity for both the words unkind and cut: unkind as in cruel and unnatural and cut as in meaning both a wound and an act that separates people. Antony is implying that Caesar was murdered not only by Brutus' cut to his body but also by Brutus' cruel cut to his soul and emotional heart. *The Final Cut* is the final version of a film shown to the audience, an idea used by Aceti (2015), who argues that managing the final cut of someone's life in a socially mediated context is deeply problematic. This is because, in death, there is a tendency to cut, edit and delete the unsavoury and in a sense rewrite someone else's history. This redaction of personal history creates new truths about someone's life. So, whilst death itself might be final physically, social media often changes the life history and the history of life. Left behind on the cutting room floor are identities, writings, artefacts and family: the biological and social immortality as mentioned in Chapter 1. To recap, biological immortality is the belief that through transmitting our genes via our descendants we continue. Social immortality is the idea that we can live on by creating artefacts that survive us, such as books, art or influences on friends or students. Yet at the same time whilst we might be asleep in an afterlife, our digital legacy continues. The difficulty with our digital legacy is not just that it continues but it also edits us out and components of our history can be edited out too. In a sense, it is our family and friends, acting as digital executors, are the cutters who choose for us which hidden remains are to be left on the cutting room floor. In this sense, our digital lives become partially deleted and managed, the product of the postdigital. We live in a postdigital world and this will have an impact on how death, dying and the afterlife are seen and managed. The postdigital is seen as a stance which merges the old and the new; it is not seen as an event or temporal position, rather it is a critical perspective, a philosophy, that can be summarized as a collection of stances in the following intersecting positions. These include disenchantment with current information systems and media, and a tendency to focus on the experiential rather than the conceptual. A persistence of digital cultures occurs that is both still digital and beyond digital, and technoscience systems continually emerge, which are transforming and transformed by global quantum computing, deep learning complexity science, artificial intelligence and big data; thus, we are not just postdigital, we are post-human as well.

The postdigital then is not just about positions or spaces inhabited just for a time – it is essentially ungraspable, as is the idea of an afterlife. Postdigital humans are located in liquid spaces; people are both central to the postdigital and key players in its formulation, interruptions and (re) creation, and this in turn overlaps with post-humanism. Post-humanism seeks to break down binary distinctions between 'human', 'machine' and 'text' and between 'nature' and 'culture'; it also rejects dualisms that are used to define being such as subject/object (Hayles, 1999 and 2012). Thus, post-humanist theory is used to question the foundational role of 'humanity'. It prompts consideration of what it means to be a human subject and the extent to which the idea of the human subject is still useful. Braidotti argues for biopower:

> not only in the sense of the government of the living but also with respect to practices of dying. By extension this means that our relationship to pain, loss and practices of mourning needs to be reconsidered in biopolitical terms.
>
> (BRAIDOTTI, 2010: 201)

Such biopower can be seen not only in the decisions made by government in the COVID-19 pandemic about who lives and who dies but also in social media platforms and those to whom we confer executorship regarding our digital bodies and therefore what constitutes the final cut. Thus the location of the dead, often seen as sleeping or waiting, results in a conundrum in relation to the digital: if the dead are sleeping, can digital afterlife have any relevance as a construct? Indeed the idea of the dead living in cyberspace may be seen by some as a nonsense, but in terms of the bereaved, it is perhaps one of the many ways that people attempt to make sense of death. The idea of the spiritual realm is spoken of in many religious texts but has become subsumed and recreated to suit society's needs. Hence whilst heaven is seen by many in society as the place where their dead loved ones reside (even if they do walk over them in a graveyard), inevitably there is little evidence that this is the case.

In the late 1990s, questions were raised about the paradox of acceleration with Virilio suggesting that 'getting closer to the 'distant' takes you away proportionally from the 'near' (and dear)' (Virilio, 1997: 80). We are in a world, in which we are, as Virilio predicted, universally telepresent without moving our physical bodies. Thus there is often a sense that we arrive and depart from spaces and encounters without actually doing so. The question then is whether we are victims or beneficiaries of this 'chronic telepresence' and there are certainly those such as Reader and Savin-Baden who argue, 'We are living through a time of acute collective ontological anxiety accentuated but not created by the digital' (2022, forthcoming). At the same time, death and the digital challenges associated with it play with the notion of temporality, introducing questions not only about the location of the dead but also about the residues of death, both physical and digital, and the impact of such residues on the living. Yet unlike Prospero in *The Tempest* (Shakespeare, 1611/2008), there is no magic to be used to time manage or count to death. Whilst Prospero's dubious practices were hidden on an island, the digital enables displays of death, whether through live-streamed funerals or open memorial pages. Remembrance and memorialization have become public and visual, with little choice about what is said or posted, as Sontag remarked:

> All photographs are memento mori. To take a photograph is to participate in another person's (or thing's) mortality, vulnerability, mutability. Precisely by slicing out this moment and freezing it, all photographs testify to time's relentless melt.
>
> (SONTAG, 1990: 15)

This relentless melt is perhaps less of a melt in the digital age, when instead we appear to have relentless remains. The consequence of this is that the accommodations around death and bereavement have changed, resulting in a digital death pragmeme and a confusion about the nature of absent presence.

ABSENT PRESENCE

The idea of absent presence originates in post-structuralist theory and is associated with Derrida (1997). The idea originally referred to the absence of the writer from a (circulated) text. Today, absent presence is a feature of all mediated communication, where people are separated from one another by time and space, as in the following example:

> A few months ago, I was out for lunch in a café bar. A family arrived: two women, one man, a teenage boy and a young girl. The young people chatted and engaged with everyone and as the food arrived, a message appeared. The man looked upset, responded to the message and was silent for much of the meal. What at first glance appeared to be a pleasant lunch out seemed to be more a case of absent presence. The most intriguing part was when the waiter asked if the meal had been okay as the group left. The response from the man was, 'Yes it was great. I just wish the dead didn't keep interrupting the living'.

The need to respond all the time, even to posthumous posts, suggests either that people control us (we have time to take the call) or that our time is more important and valuable than those we are with face-to-face (we take the call to show how precious our time is). We live in a world where the present and absent have an equivalent value

The practice of virtual archiving ensures a relationship with the deceased that reflects the idea of both their presence and absence. Social media sites are used in ways that ensure that mourning is not based on death and loss but instead upon the replication of a social media's never-ending life, an absent presence. Today experiences of absence and presence have become more prevalent due to the ever-increasing use of social media, as demonstrated in Table 12.1.

Table 12.1 presents a binary position, and philosophers such as Derrida (1997) have suggested that languages, images and ways of representation can help us to see that social media capture our presence. For example, a number of authors describe presence as the sense of being 'in' or 'part of' a 3D virtual world. In 3D worlds, a user interacts with people, objects and places in these virtual worlds using an avatar. An avatar is a virtual presence that may or may not be representative of a user's real-life identity. An avatar will have a name and, in most (though not necessarily all) instances, a distinct appearance. There are a number of virtual worlds in operation, all with very different approaches to creating 'realities'. Some are photo-realistic, others impressionistic; some focus on fantasy, others reality, but most have an emphasis on the interaction. Notably in the film *The Matrix*, the question is asked 'what is real?'

TABLE 12.1   Absence and Presence

Description	Social media example	Digital afterlife example
Present presence: talking to someone face-to-face	Speaking face-to-face via Skype	Speaking to someone who is dead on the phone as if they were still there
Present absence: speaking to someone directly who is not with us	Speaking to someone directly on the phone	Messaging the dead on Facebook
Absent presence: someone you are with is communicating with someone elsewhere	You are talking to someone face-to-face and they pull out their phone to post a message on Instagram	Speaking to the post-death avatar of your loved one
Absent absence: the complete absence of the person	Online shrines or media-enhanced headstones	Using spaces of digital remembrance to recognize absence such as virtual graveyards

Right now are we just inside a computer program?

Your appearance now is what we call residual self-image. It is the mental projection of your digital self

This isn't real?

What is real? How do you define real? If you were talking about what you can feel, what you can smell, and taste and see, then real is simply electrical signals interpreted by your brain.

(WACHOWSKI AND WACHOWSKI, 1999)

The reanimation of digital identities begins to introduce questions about what counts as being real. The consequence of this is the suggestion of the possibility of cheating death by merging real and imagined lives. Further, the reanimation of people after death has, to some extent, collapsed death by creating an illusion, as Aceti argues:

> By constantly digitising and uploading data/lives, 'digital human-ity' finds and confirms its existence as an eternal illusion and an eternal reality that constantly attempt to grasp themselves via a screen in the opposition of real vs. virtual.

(ACETI, 2015: 321)

Yet, despite this illusion, context collapse (boyd, 2002: 12) is rarely mentioned in relation to digital afterlife; it as is if such collapse has not come to the fore in death-online studies, instead life-death-digital context collapse has just become woven into every day existence. However, despite context collapse symbolic behaviours remain; and these do seem to transcend both physical and digital behaviours in relation to death and mourning, whether language or conventions.

THE DIGITAL DEATH PRAGMEME

There is a sense that social media is not only an interruption of life but an interruption of funeral practices, ways of mourning and views of death and the afterlife. The result is that for some people, the value of the ways someone's death is mourned can be measured by the posts, responses to posts and the perspectives offered about the post by those who never knew the person. The nature of the language used to accommodate someone's death face-to-face and the practices and conventions are, in the main, well known and understood within given cultures. There are conventions around the period of dying, the event of death and post-death. Acceptable practices include weeping over the dead body, eulogizing over a coffin and sending sympathy cards. What is interesting is that Parvaresh (2017) argues that linguistic devices and behaviours constitute a pragmeme of accommodation. A pragmeme (Mey, 2006) is a situated speech act which is not dependent on the words themselves but on the context and culture in which it is situated. Thus, in relation to the pragmeme of accommodation in death, the importance is in the construction of meaning in relation to the interaction that is taking place:

> The interaction involved in the mourning practices are, in fact, subordinated to giving solace to the close relatives of the deceased and to ensure that they *accommodate* to the new state of affairs.
>
> (CAPONE, 2010: 6; italics added)

Using the term 'pragmeme of accommodation' helps to understand the language rituals and the practices used in the context of dying and death. What is both troublesome and complex is that the pragmeme of accommodation tends to be culturally embedded. Thus, in digital spaces accommodation mistakes are often made. Digital spaces are invariably used to try to mimic real-life spaces, but in terms of language, rituals and practices related to dying and death, this often fails. What occurs instead is that people break away from the pragmeme of accommodation and speak directly to the dead, using a linguistic style which suggests affection and a close relationship with the deceased even if this is not the case, as Giaxoglou notes:

> Most notably, conventional funerary expressions appearing at important discourse junctures of the message combine with turns of everyday talk, such as greetings and leave-takings, terms of address and endearment and construct a sense of unbroken post-mortem relationality and bonds with the deceased.
>
> (GIAXOGLOU, 2014: 166)

What is needed for online death spaces is a deeper awareness of a digital death pragmeme; the idea that situated speech acts relating to death and dying in digital spaces should be guided by both the context and culture in which they are placed. This should be in terms of reflecting both the culture of the physical location of the deceased and the netiquette of online memorialization.

DIGITAL AFTERLIFE AND THE SPIRITUAL REALM

The idea of the spiritual realm is something that seems decidedly alien in the 21st-century postdigital world. Few people seem to speak of the spiritual, even less of the spiritual realm. At the same time, discussions about the afterlife are ones that constantly seemed to stall people; they perhaps do not want to consider their own death or to contemplate where they might go after death. The prevalence of death in social media spaces might therefore be an attempt to obfuscate death. Thus, people appear to use social media spaces as a means of avoidance or even escape. It is also clear but at the same time troublesome that few people wish to discuss spiritual issues; perhaps spiritual issues like death are taboo, whilst mindfulness and spiritualism are seen as acceptable. At the same time, there seems to be a shift in the nature of the spiritual and the sacred. At one level, this appears to be the professionalization of the sacred, so that funeral events are becoming and treated like a film set, with funeral directors parading as chat show hosts and with cameras interrupting the ritual, ceremony and the ability to grieve. On the other hand, there is a perception that pixels themselves can be seen as sacred ground (Kuipers, 2021). Thus when glitches and problems occur on zoom or streamed funerals, this seems troublesome and deeply problematic, which appears to result in confusion about the relationship between ritual, the spiritual, the sacred and the management of death.

It is evident that despite the range of research, articles and conferences on death online as well as grief festivals, relatively few people engage in-depth with the spiritual component of death. This might be because it is easier to consider a digital will, digital assets and digital remains as something far-removed, or because it is due to a denial of one's own mortality. This seems to raise questions about how many people actually have a Facebook legacy contact, how people who work in the digital arena have curated and saved their digital assets offline, and whether they have created a realistic digital will. It is also apparent that there remains considerable legal difficulty in accessing people's digital remains and to date, many people ignore the law and use the deceased's password to access their digital assets. It seems that what is needed is a rethinking of property law and clear advice to guide people about archiving their digital assets in realistic ways.

What is interesting at the end of this book is that to reflect on the idea that whilst there may be physical death, there can be no end to digital media and thus, in fact, no real digital death; what we seem to have is digital eternity. In the 21st century, I would suggest that it is in fact no longer possible to be *not* digital anymore, even if attempts are made to become 'willfully unconnected' (Hanson, 2013: 224). Hanson suggested that the wilfully unconnected are those who have chosen not to use digital technologies, making a conscious choice to marginalize themselves from the Information Society. She terms these as the *new minority* and suggests that this new minority reminds us that using technology is a choice and not a requirement for our lives in a contemporary society. Yet, at the same time, it would seem that many of us who choose to engage (often) in digital and social media spaces are creating an unhelpful amount of media waste, as Lagerkvist suggests:

> In our age of ecological crisis, media are also increasingly a seeming wasteland, an end station, a terminus, a final depot, as we are in the anthropocene faced with enormous amounts of media waste, and with the environmental problems of the materiality of media death. Hence our new media age is a life among the dying media, where we wade through media junk and outmoded, obsolete and ageing apparatuses as well as abandoned platforms and other atrophied forms of media life.
>
> (LAGERKVIST, 2017: 56)

Perhaps what we are really discussing is some kind of digital eternity rather than an understanding of the spiritual realm, the idea that everything can be saved and stored in some kind of omnipotent archive. However, it is clear from recent studies that there are still those who believe in literal immortality; the idea of an afterlife which is based on the belief of immortality of the soul, which whilst largely attributed to those with religious faith, is not exclusively the case (Georgiadou and Pnevmatikos, 2019). Furthermore, when exploring perceptions of the afterlife in relation to digital media it is clear that those who speak to the dead on social media sites do appear to believe in a form of literal immortality (Kasket, 2019; O'Connor and Kasket, 2021). Whilst the afterlife in the main is still largely seen to be connected to religious belief, it is clear that in the context of social media there is still a transposition across both beliefs and media.

Thus whilst people in social media spaces may not use the term 'afterlife', they do imply or even believe that a biological death is not the end of existence. This would suggest there is mounting credence in the belief of a digital afterlife, which is possibly a result of the living needing to maintain an ongoing relationship with the deceased, despite a lack of any personal clarity about their own belief in any kind of afterlife.

REFLECTION

It could be argued that there is a need to acknowledge the left behind dead as media junk, similar to the notion of space junk: human-made junk left in space whether satellite or debris. Such media junk introduces not only questions about environmental impact but also the relative value of the digital dead living on beyond the next digital dead. A digital afterlife full of media junk seems to have little purpose or appeal. As mentioned in Chapter 1, the exploration of the afterlife introduces questions about what it means to be human, and Harari (2015) challenges us to consider whether indeed there is a next stage of evolution and asks where do we go from here? Yet our purpose, our very existence has always been deeply troublesome to humanity, reflected best perhaps by the existentialists and the novels of Camus in *L'Etranger* (Camus, 1942a) and *Le Mythe de Sisyphe* (Camus, 1942b). However, perhaps there is hope, perhaps the world will not end with a whimper as T.S. Eliot suggested at the end of *The Wasteland* (Eliot 1922/1999). Perhaps the afterlife will be as C.S. Lewis suggested:

> It is as hard to explain how this sunlit land was different from the old Narnia as it would be to tell you how the fruits of that country taste. Perhaps you will get some idea of it if you think like this. You may have been in a room in which there was a window that looked out on a lovely bay of the sea or a green valley that wound away among mountains. And in the wall of that room opposite to the window there may have been a looking-glass. And as you turned away from the window you suddenly caught sight of that sea or that valley, all over again, in the looking glass. And the sea in the mirror, or the valley in the mirror, were in one sense just the same as the real ones: yet at the same time there were somehow different — deeper, more wonderful, more like places in a story: in a story, you have never heard but very much want to know. The difference between the old Narnia and the new Narnia was like that. The new one was a deeper country: every rock and flower

and blade of grass looked as if it meant more. I can't describe it any better than that: if ever you get there you will know what I mean. It was the Unicorn who summed up what everyone was feeling. He stamped his right fore-hoof on the ground and neighed, and then he cried: 'I have come home at last! This is my real country! I belong here. This is the land I have been looking for all my life, though I never knew it till now'.

(LEWIS, 1956: 209)

References

Aceti, L. (2015). Eternally present and eternally absent: The cultural politics of a thanatophobic Internet and its visual representations of artificial existences. *Mortality*, *20*(4), 319–333, doi:10.1080/13576275.2015.1085297

Alves, R. V. (2021). The Internet, social media and grieving. *Médecine Palliative*, *20*(3), 155–158.

Andersen, C. U., Cox, G., & Papadopoulos, G. (2014). Postdigital research. *A Peer-Reviewed Journal About*, *3*(1). Available at https://aprja.net//issue/view/8400

Ariès, P. (1974). *Western Attitudes Toward Death from the Middle Ages to the Present*. Baltimore: Johns Hopkins University Press.

Bainbridge, W. (2013). Perspectives on virtual veneration. *Information Society*, *29*(3), 196–202.

Bainbridge, W. (2017). The virtual conquest of death. In M. H. Jacobsen (Ed.), *Postmortal Society – Towards a Sociology of Immortality*. London: Routledge.

Barad, K. (2003). Posthumanist performativity: Toward an understanding of how matter comes to matter. *Signs: Journal of Women in Culture and Society*, *28*(3), 801–831. doi:10.1086/345321.

Barberia, I., Oliva, R., Bourdin, P., & Slater, M. (2018). Virtual mortality and near-death experience after a prolonged exposure in a shared virtual reality may lead to positive life-attitude changes. *PLoS ONE*, *13*(11), e0203358. doi:10.1371/journal.pone.0203358

Barclay, W. (1998). *The Apostles Creed*. London: John Knox Press.

Bassett, D. J. (2020). Profit and loss: The mortality of the digital immortality platform. In M. Savin-Baden & V. Mason-Robbie (Eds.), *Digital Afterlife. Death Matters in a Digital Age*. Boca Raton, FL: CRC Press, 75–88.

Bassett, D. J. (2018a). Ctrl+Alt+Delete: The changing landscape of the uncanny valley and the fear of second loss. *Current Psychology*, 1–9. doi:10.1007/s12144-018-0006-5

Bassett, D. (2018b). Digital afterlives: From social media platforms to Thanabots and beyond. In C. Tandy (Ed.), *Death and Anti-Death, Vol. 16: 200 Years After Frankenstein*. Ann Arbor: Ria UP.

Beck, U. (1992). *Risk Society*. London: Sage.

Becker, E. (1973). *The Denial of Death*. New York: Free Press.

Bell, J., & Bailey, L. (2017). The use of social media in the aftermath of a suicide: Findings from a qualitative study in England. In T. Niederkrotenthaler & S. Stack (Eds.), *Media and Suicide International Perspectives on Research, Theory, and Policy*, 75–86. Abingdon, Oxon: Taylor & Francis.

Bell, G., & Gray, J. (2000). *Digital Immortality*. Technical report for Microsoft Research. Available at https://www.microsoft.com/en-us/research/wp-content/uploads/2016/02/tr-2000-101.pdf.

Berry, D. M., & Fagerjord, A. (2017). *Digital Humanities: Knowledge and Critique in a Digital Age*. Cambridge: Polity Press.

Birnhack, M., & Morse, T. (2018). *Regulating Access to Digital Remains – Research and Policy Report*. Israeli Internet Association. Available at https://www.isoc.org.il/wp-content/uploads/2018/07/digital-remains-ENG-for-ISOC-07-2018.pdf.

Blake, W. (1794). *The Marriage of Heaven and Hell*. New York: Dover Publications Inc.

Blondheinm, M., & Rosenburg, H. (2017). Media theology: New Communication Technologies as religious constructs, metaphors, and experiences. *New Media & Society, 19*(1), 43–51.

Bosetti, L., Kawalilak, C., & Peggy Patterson, P. (2008). Betwixt and between: Academic women in transition. *Canadian Journal of Higher Education, 38*(2), 95–115.

boyd, d. (2002). *FACETED ID/ENTITY: Managing Representation in a Digital World*, Master of Science in Media Arts and Sciences, Massachusetts Institute of Technology, September 2002. Available at www.danah.org/papers/Thesis. FacetedIdentity.pdf

Bragg, M. (2007). We tend to forget that life can only be defined in the present tense'. Edited version of Melvyn Bragg's interview of Dennis Potter on March 15 1994. It was broadcast by Channel 4 on April 5 1994. Available at https://www.theguardian.com/theguardian/2007/sep/12/greatinterviews

Braidotti, R. (2010). The politics of 'life itself and new ways of dying. In D. Coole & S. Frost (Eds.), *New Materialisms: Ontology, Agency, and Politics*. Durham: Duke University Press, 201–218.

Brittz, K. (2018). Soul searching: Finding space for the soul in the New Digital Age. *Image and Text, 31*, 1–29.

Brubaker, J. R., Hayes, G. R., & Mazmanian, M. (2019). Orienting to networked grief: Situated Perspectives of communal mourning on Facebook. *Proceedings of the ACM on Human-Computer Interaction, 3*, CSCW, Article 27 (November 2019), 19 pages. doi:10.1145/3359129Brubaker

Brubaker, J. R., Hayes, G. R., & Dourish, P. (2013). Beyond the grave: Facebook as a site for the expansion of death and mourning. *The Information Society, 29*(3), 152–163, doi: 10.1080/01972243.2013.777300

Brubaker, J. R, & Vertesi, J. (2010). Death and the social network. In HCI at the end of life workshop at CHI2010, Savannah, GA. Available at https://pdfs.semanticscholar.org/aaf3/b480dc877adde04ab57421aef59aaee8b4f3.pdf

Bruner, J. S. (1990). *Acts of Meaning*. Cambridge, MA: Harvard University Press.

Bruner, J. S. (2002). *Making Stories: Law. Literature, Life*. New York: Farrar, Straus and Giroux.

Bruns, A., & Burgess, J. (2015). Twitter hashtags from ad hoc to calculated publics. In N. Rambukkana (Ed.), *Hashtag Publics: The Power and Politics of Discursive Networks [Digital Formations, Volume 103]*. NewYork: Peter Lang Publishing Group, 13–27.

Burden, D. (2020). Building a digital immortal. In M. Savin-Baden & V. Mason-Robbie (Eds.), *Digital Afterlife. Death Matters in a Digital Age*. Boca Raton, FL: CRC Press.

Burden, D., & Savin-Baden, M. (2019). *Virtual Humans: Today and Tomorrow*. Boca Raton, FL: CRC Press.

Campbell, H. A. (2012a). Understanding the relationship between religion online and offline in a networked society. *Journal of the American Academy of Religion*, *80*(1), 64–93. doi:10.1093/jaarel/lfr074.

Campbell, H. A. (Ed.). (2012b). *Digital Religion: Understanding Religious Practice in New Media Worlds*. New York: Routledge.

Campbell, H. A., & Golan, O. (2011). Creating digital enclaves: Negotiation of the internet among bounded religious communities. *Media, Culture & Society*, *33*(5), 709–724.

Campbell, H. A., & Garner, S. (2016). *Networked Theology. Negotiating Faith in Digital Culture*. Grand Rapids, MI: Baker Academic.

Campbell, H. (2017). Religious communication and technology. *Annals of the International Communication Association*, *41*, 1–7. doi:10.1080/23808985.2 017.1374200.

Campbell H.A., & Rule F. (2020). The practice of digital religion. In H. Friese, M. Nolden, G. Rebane, & M. Schreiter (Eds.), *Handbuch Soziale Praktiken und Digitale Alltagswelten*. Wiesbaden: Springer VS.

Camus, A. (1942a). *L'Entrange* . Paris: Gallimard.

Camus, A. (1942b). *Le Mythe de Sisyphe*. Paris: Gallimard.

Cann, C. (2018). Buying an Afterlife: mapping religious beliefs through consumer death goods. In C. Cann (Ed.), *The Routledge Handbook of Death and Afterlife*. Oxford: Routledge.

Cann, C. (2014). Tweeting death, posting photos, and pinning memorials: Remembering the dead in bits and pieces. In C. M. Moreman & A. D. Lewis (Eds.), *Digital Death Mortality and Beyond in the Online Age*. Santa Barbara, CA: Praeger.

Capone, A. (2010). On pragmemes again. Dealing with death. *La linguistique*, *46*, 3–21.

Cellan-Jones (2014). Stephen Hawking warns artificial intelligence could end mankind. *BBC News* (2 December, 2014). Available at https://www.bbc.co.u k/news/technology-30290540

Chapman, B. (2018). Nearly two-thirds of adult don't have will, research finds. *Independent*. January 9. Available at https://www.independent.co.uk/news/ business/news/nearly-two-thirds-ukadults-don-t-have-will-research-finds-a8148316.html.

Clabburn, O., Groves, K.E., & Jack, B. (2020). Virtual learning environment ('Ivy Street') for palliative medicine education: Student and facilitator evaluation. *BMJ Supportive & Palliative Care, 10*, 318–323.

Clarke, J. N. (2006). Death under control: The portrayal of death in mass print English Language Magazines in Canada. *OMEGA, 52*(2), 153–167.

Clandinin, D. J., & Connelly, F. M. (2000). *Narrative Inquiry*. San Francisco, CA: Jossey-Bass.

Clayden, A., Green, T., Hockley, J., & Powell, M. (2010). From cabbages to cadavers: natural burial down on the farm. In Maddrell & J. D. Didaway (Eds.), *Deathscapes. Spaces for Death, Dying, Mourning and Remembrance*. Oxfordshire: Routledge, 95–118.

Collopy, B. J. (1978). Theology and the darkness of death. *Theological Studies, 39*(1), 22–54.

Cook, S. L. (2007). Funerary practices and afterlife expectations in ancient Israel. *Religion Compass, 1*(6), 660–683.

Cook, D. M., Dissanayake, D.N., & Kaur, K. (2019). The usability factors of lost digital legacy data from regulatory misconduct: Older values and the issue of ownership. In 2019 7th International Conference on Information and Communication Technology (ICoICT), 1–6.

Cook-Sather, A., & Alter, Z. (2011). What is and what can be: How a liminal position can change learning and teaching in higher education. *Anthropology & Education Quarterly, 42*(1), 37–53.

Cramer, F. (2015). What is 'post-digital'? In D. M. Berry & M. Dieter (Eds.), *Postdigital Aesthetics: Art, Computation and Design* (pp. 12–26). New York: Palgrave Macmillan. doi:10.1057/9781137437204.

Cuminskey, K., & Hjorth, L. (2018). *Haunting Hands*. Oxford, UK: Oxford University Press.

Daer, A. R., Hoffman A. F., & Goodman, S. (2014). Rhetorical functions of hashtag forms across social media applications. *Communication Design Quarterly, 3*(1), November. Available at http://www.users.miamioh.edu/simmonwm/hashtags_daer.pdf

Data Protection Act 2018, *c*. 12. Available at http://www.legislation.gov.uk/ukpga/2018/12/contents/enacted

Davies, D. J. (2008). *The Theology of Death*. London: Continuum.

Day, A. (2011). *Believing in Belonging*. Oxford: Oxford University Press.

de Certeau, M. (1984). *The Practice of Everyday Life*. (S. Rendall trans.). Berkeley, CA: University of California Press.

Deering, B. (2010). From Anti social behaviour to X-rated: Exploring social diversity and conflict in the cemetery. In A. Maddrell & J. D. Didaway (Eds.), *Deathscapes. Spaces for Death, Dying, Mourning and Remembrance*. Oxfordshire: Routledge, 73–93.

DeGroot, J. M. (2014). 'For whom the bell tolls': Emotional rubbernecking in Facebook memorial groups. *Death Studies, 38*(2), 79–84. doi:10.1080/0748 1187.2012.725450.

Deleuze, G. (1993). *The Fold: Leibniz and the Baroque*. (T. Conley trans.). New York: Continuum.

Derrida, J. (1997). *Of Grammatology.* (G. Spivak trans.). Baltimore: Johns Hopkins University Press.

Derrida, J., & Vattimo, G. (1998). *Religion.* Cambridge: Polity Press.

Dickens, C. (1843). *A Christmas Carol.* London: Chapman & Hall.

Doka, K. (2008). Disenfranchised grief in historical perspective. In M. S. Stroebe, R. O. Hansson, H. Schut, & W. Stroebe (Eds.), *Handbook of Bereavement Research and Practices: Advances in Theory and Intervention.* Washington, DC: American Psychological Association, 223–240.

Drescher, E. (2012). Pixels perpetual shine: The mediation of illness, dying, and death in the digital age. *CrossCurrents, 62*(2), 204–218. doi:10.1111/j.1939-3881.2012.00230.x

du Preez, A. (2018). Sublime selfies: To witness death. *European Journal of Cultural Studies, 21*(6), 744– 760.

Durkheim, E. (1893). *The division of labour in society.* (G. Simpson trans.). Available at https://archive.org/details/in.ernet.dli.2015.126617

Eck, D. L. (1998). *Darsan: Seeing the Divine Image in India.* New York: Columbia University Press.

Eliot, T. S. (1922/1999). *The Wasteland and Other Poems.* London: Faber and Faber.

Eter9 (2017). Available at https://www.eter9.com/

Eternime (2017). Available at http://eterni.me/

Ferré, J. P. (2003). The media of popular piety. In S. M. Jolyon & P. Mitchell (Eds.), *Mediating Religion: Studies in Media, Religion, and Culture.* London; New York: T&T Clark, pp. 83–94.

Fitzpatrick, B., & Recordon, D. (2007). Thoughts on the social graph. Online. Available at http://bradfitz.com/social-graph-problem/

Foley, M., & Lennon, J. (1996). JFK and dark tourism: A fascination with assassination. *International Journal of Heritage Studies, 2,* 98–211.

Foltyn, J. L. (2021). Touring heaven and hell: Spectacular encounters by celebrities in near-death experiences. In M. H. Jacobsen (Ed.), *The Age of Spectacular Death.* London: Routledge, 52–73.

Foucault, M. (1987). *Death and the Labyrinth: The World of Raymond Roussel.* (C. Ruas trans.). London: Athlone Press.

Freud, S. (1918/1963). Refection on war and death. In P. Rieff (Ed.), *Character and Culture.* New York: Collier Books, 107–134.

Fullerton, S. (2020).What else is there? *Oneing, 8*(1), 77–80.

Gach, K. Z., Fiesler, C. & Brubaker, J. R. (2017). "Control your emotions, Potter:" An analysis of grief policing on Facebook in response to celebrity death. SIG Paper in Word Format. ACM Trans. Web, *1,* 2, Article 47 (November 2017), 13 pages. doi:10.1145/3134682. Available at https://www.research gate.net/publication/321606129_Control_your_emotions_Potter_ An_Analysis_of_Grief_Policing_on_Facebook_in_Response_to_Cel ebrity_Death

Georgiadou, T., & Pnevmatikos, D. (2019). An exploration of afterlife beliefs in religiously-and secularly-oriented adults. *Journal of Beliefs & Values, 40*(2), 159–171, doi: 10.1080/13617672.2019.1583921

Giaxoglou, K. (2014). Language and affect in digital media: Articulations of grief in online spaces for mourning. In B. O'Rourke, N. Bermingham, & S. Brennan (Eds.), *Opening New Lines of Communication in Applied Linguistics: Proceedings of the 46th Annual Meeting of the British Association for Applied Linguistics*. London: Scitsiugnil Press, 161–171.

Giaxoglou, K. (2020). Mobilizing grief and remembrance with and for networked publics: towards a typology of hyper-mourning. *European Journal of Life Writing*, 9(2), 264–284. doi: 10.21827/ejlw.9.36910

Gibson, J. (1979). *The Ecological Approach to Visual Perception*. Boston, MA: Houghton Mifflin.

Gibson, M. (2019). Death in *Second Life*: Lost and missing lives. In T. Kohn, M. Gibbs, B. Nanasen, & L. van Run (Eds.), *Residues of Death, Disposal Refigured*. Abingdon: Routledge.

Gittings, C., & Walter, T. (2010). Rest in peace? Burial on private land. In Maddrell & J. D. Didaway (Eds.), *Deathscapes. Spaces for Death, Dying, Mourning and Remembrance*. Oxfordshire: Routledge, 95–118.

Gomes de Andrade, N., Pawson, D., & Muriello, D. et al. (2018). Ethics and Artificial Intelligence: Suicide prevention on Facebook. *Philosophy Technology*, 31, 669–684. doi:10.1007/s13347-018-0336-0

Gooder, P. (2011). *Heaven*. London: SPCK Publishing.

Gould, H., Arnold, M., Dupleix, T., & Kohn (2021). 'Stood to rest': Reorientating necrogeographies for the 21st century, *Mortality*. doi:10.1080/13576275.202 1.1878120

Gustavsson, A. (2011). Conceptions of faith as expressed on memorial internet websites in Norway and Sweden. An existence after death? In A. Gustavsson (Ed.), *Cultural Studies on Death and Dying in Scandinavia*. Oslo, Norway: Novus, 142–161.

Hallam, E., & Hockey, J. (2001). *Death, Memory and Material Culture*. Oxford: Berg.

Hamonic, W. G. (2017). Global Catastrophe in motion pictures as meaning and message: The functions of Apocalyptic cinema in American film. *Journal of Religion & Film*, 21(1). Article 36. Available at https://digitalcommons.un omaha.edu/jrf/vol21/iss1/36

Han, N. (2015, May 1). Donor beware: Concerns about crowdfunding campaigns. *ABC News*. Available at http://6abc.com/finance/donor-beware-concern-about- crowdfundingsites/ 690073/

Hanson, J. (2013). The new minority: the Willfully unconnected. In P. M. A. Baker, J. Hanson & J. Hunsinger (Eds.), *The Unconnected: Social Justice, Participation, and Engagement in the Information Society*. New York: Peter Lang, 223–240.

Harari, Y. N. (2015). *Homo Deus*. London: Harvill Secker.

Harbinja, E. (2017a). *Legal Aspects of Transmission of Digital Assets on Death* (Unpublished PhD thesis). University of Strathclyde.

Harbinja, E. (2017b). Post-mortem privacy 2.0: Theory, law, and technology. *International Review of Law, Computers & Technology*, 31(1), 26–42. doi:10. 1080/13600869.2017.1275116

Harbinger, E. (2020). The 'New(ish)' Property, Informational Bodies, and Postmortality. In M. Savin-Baden & V. Mason-Robbie (Eds.), (2020). *Digital Afterlife. Death Matters in a Digital Age*. Boca Raton, FL: CRC Press.

Hård af Segerstad, Y., Bell, J., & Yeshua-Katz, D. (2020). Designed to die: on The ephemerality and obsolesce of digital remain in Hård af Segerstad, Y., Bell, J., Giaxoglou, K., Pitsillides, S., Yeshua-Katz, D., Cumiskey, K., & Hjorth, L. Taboo or not taboo: (in)visibilities of death, dying and bereavement. *AoIR Selected Papers of Internet Research*, 2020. doi:10.5210/spir.v2020i0.11125

Hayles, K. (1999). *How We Became Posthuman: Virtual Bodies in Cybernetics, Literature and Informatics*. Chicago, IL: University of Chicago Press.

Hayles, N. K. (2012). *How We Think: Digital Media and Contemporary Technogenesis*. Chicago, IL: University of Chicago Press.

Healy, C. (2016). Thin place: An alternative approach to curatorial practice. A thesis submitted in partial fulfilment of the requirements of the University of the West of England, Bristol for the degree of PhD.

Helland, C. (2012). Scholar's Top 5: Christopher Helland on online religion and religion online. NNMRDC Blog Series. Available at https://digitalreligion .tamu.edu/blog/mon-05142012-1132/scholar's-top-5-christopher-helland-o nline-religionand-religion-online

Hick, J. (2004). *The Fifth Dimension: An Exploration of the Spiritual Realm*. Oxford, England: Oneworld Publications.

Hjarvard, S. (2008). The mediatization of religion: A theory of the media as agents of religious change. *Nordic Journal of Media Studies*, 6, 9–26. doi:10.1386/ nl.6.1.9_1

Hjorth, L., & Cumiskey, K. M. (2019). Selfie eulogies. In T. Kohn, M. Gibbs, B. Nansen, & L. van Ryn (Eds.), *Residues of Death Disposal Refigured*. London: Routledge.

Hjorth, L., & Cumiskey, K. M. (2018). Mobiles facing death: Affective witnessing and the intimate companionship of devices. *Cultural Studies Review*, 24(2), 166–180.

Hockey, J. (1996). Encountering the 'reality of death' through professional discourses: The matter of materiality. *Mortality*, 1, 45–61.

Hodalska, M. (2019). Selfies at horror sites: Dark tourism, ghoulish souvenirs and digital narcissism. *Zeszyty PRASOZNAWCZE*, 230(2), 405–423.

Homer, Fagles, R., & Knox, B. (1998). *The Iliad*. New York: Penguin Books.

Howarth, G. (2000). Dismantling the boundaries between life and death. *Mortality*, 5, 127–138.

Hutchings, T. (2019). Angels and the Digital Afterlife: Death and Nonreligion Online. *Secularism and Nonreligion*, 8(7), 1–6. doi:10.5334/snr.105

Ibbetson, C. (2019). What should happen to data and social media accounts when people die? *YouGov*, November 1. Available at https://yougov.co.uk/topics/li festyle/articles-reports/2019/11/01/what-do-brits-want-happen-their-data-and-social-me

Ito, M., Baumer, S., Bittanti, M., boyd, d., Cody, R., Herr-Stephenson, B. et al. (2010). *Hanging Out, Messing Around, and Geeking Out*. Cambridge, MA: MIT Press.

Jacobsen, M. H. (2021). Introduction. In M. H. Jacobsen (Ed.), *The Age of Spectacular Death*. London: Routledge.

Jacobsen, M. H. (2017). 'The Bad Death': Deciphering and developing the dominant discourse on 'The Good Death'. In V. Parvaresh & A. Capone (Eds.), *The Pragmeme of Accommodation: The Case of Interaction Around the Event of Death*. Perspectives in Pragmatics, Philosophy & Psychology, (Vol. *13*). Cham: Springer. doi:10.1007/978-3-319-55759-5_18

Jandrić, P. (2019). The postdigital challenge of critical media literacy. *The International Journal of Critical Media Literacy*, *1*(1), 26–37. doi:10.1163/25900110-00101002.

Jandrić, P., Knox, J., Besley, T., Ryberg, T., Suoranta, J., & Hayes, S. (2018). Postdigital science and education. *Educational Philosophy and Theory*, *50*(10), 893–899. doi:10.1080/00131857.2018.1454000.

Jaspers, K. (1953). *The Origin and Goal of History*. New Haven, CT and London: Yale University Press.

Johnson, P. (2013). The geographies of heterotopia. *Geography Compass*, *7*(1), 790–803.

Kania-Lundholm, M. (2019). Digital mourning labor: Corporate use of dead celebrities on social media. In T. Holmbery, A. Jonsson, & F. Palm (Eds.), *Death Matters. Cultural Sociology of Mortal Life*. Cham: Palgrave Macmillan.

Kant, E. (1781/2007). *Critique of Pure Reason* (Rev. edition). 29 November 2007. London: Penguin Classics.

Kasket, E. (2021). If death is the spectacle, big teach is the lens: How social media frame an age of 'spectacular death'. In M. H. Jacobsen (Ed.), *The Age of Spectacular Death*. Oxford: Routledge.

Kasket, E. (2019). *All the Ghosts in the Machine: Illusions of Immortality in the Digital Age*. London, UK: Robinson.

Kasket, E. (2012). Continuing bonds in the age of social networking: Facebook as a modern-day medium. *Bereavement Care*, *31*(2), 62–69.

Kastenbaum, R. J. (2004). *Death, Society, and Human Experience* (8th edition). Boston, MA: Allyn & Bacon.

Kearl, M. C. (2021). The proliferation of skulls in popular culture. In M. H. Jacobsen (Ed.), *The Age of Spectacular Death*. London: Routledge, 74–88.

Keefer, L. A., Landau, M. J., Rothschild, Z. K., & Sullivan, D. (2012). Attachment to objects as compensation for close others' perceived unreliability. *Journal of Experimental Social Psychology*, *48*(4), 912–917. doi:10.1016/j.jesp.2012.02.007

Kellaher, L., & Worpole, K. (2010). Bringing the dead back home: Urban public spaces as sites for new patterns of mourning and memorialisation. In A. Maddrell & J. D. Sidaway (Eds.), *Deathscapes* (pp. 161–180). Farnham: Ashgate.

Kember, S. & Zylinska, J. (2011). *Life After New Media: Mediation as a Vital Process*. Cambridge, MA: MIT Press.

Khapaeva, D. (2021). Killing humanity. Anthropocentrism and apocalypse in contemporary film. In M. H. Jacobsen (Ed.), *The Age of Spectacular Death*. Oxford: Routledge.

Klass, D. (2018). Prologue. In D. Klass & E. M. Steffen (Eds.), *Continuing Bonds in Bereavement: New Directions for Research and Practice.* New York: Routledge, xiii–xix.

Klass, D., & Heath, A. O. (1996–1997). Grief and abortion: *Mizuko Kuyo,* the Japanese ritual resolution. *Omega: Journal of Death and Dying, 34*(1), 1–14.

Klass, D., Silverman, P. R., & Nickman, S. L. (Eds.). (1996). *Continuing Bonds: New Understandings of Grief.* Washington, DC: Taylor & Francis.

Klastrup, L. (2015). "I didn't know her, but…": Parasocial mourning of mediated deaths on Facebook RIP pages. *New Review of Hypermedia and Multimedia, 21*(1–2), 146–164. doi:10.1080/13614568.2014.983564

Kneese, T. (2018). Mourning the commons: Circulating affect in crowd-funded funeral campaigns. *Social Media + Society,* 1–12. doi:10.1177/2056 305117743350

Kofoed, J. (2008). Muted transitions. *European Journal of Psychology of Education, 23*(2), 199–212.

Kollowe, J. (2018). UK funerals industry under investigation for high prices. Available at https://www.theguardian.com/business/2018/nov/29/uk-fun erals-industry-under-investigation-high-prices-cma

Kubler-Ross, E. (1969). *On Death and Dying.* London: Macmillan.

Kuipers, J. (2021). Talking about funerals: How participants in COVID funerals reveal their . . . Paper presented at Death Online Research Symposium IT University of Denmark 21–23 April.

Kurzweil, R. (1999). *The Age of Spiritual Machines: When Computers Exceed Human Intelligence.* Knutsford, UK: Texere Publishing.

Lagerkvist, A. (2017). The media end: Digital afterlife agencies and techno-existential closure. In A. Hoskins (Ed.), *Digital Memory Studies: Media Pasts in Transition.* New York: Routledge, 48–84.

Lagerkvist, A., & Andersson, Y. (2017). The grand interruption: Death online and mediated lifelines of shared vulnerability. *Feminist Media Studies, 17*(4), 550–564, doi: 10.1080/14680777.2017.1326554

Lammes, S. (2008). Spatial regimes of the digital playground: Cultural functions of spatial practices in computer games. *Space and Culture, 11,* 260–272.

Land, R., Rattray, J., & Vivian, P. (2014). Learning in the liminal space: A semiotic approach to threshold concepts. *Higher Education, 67,* 199–217.

Landström, M., & Mustafa, N. (2018). Developing an Artificially Intelligent tool for grief recovery. Master of Science Thesis MMK 2017: 172 IDE 304. Available at http://www.diva-portal.org/smash/get/diva2:1211163/FULLTEXT01.pdf

Lau, W. (2016). Vespers, the latest mask collection by MIT's Neri Oxman. Available at https://www.architectmagazine.com/technology/vespers-the-latest-mask-collection-by-mits-neri-oxman_o (December 15).

Lefebvre, H. (1991). *The Production of Space* (15th edition). Oxford: Blackwell.

Legislation.gov.uk. (2016). *Computer Misuse Act 1990.* [online] Available at http://www.legislation.gov.uk/ukpga/1990/18/section/3ZA

Legislation.gov.uk. (2015). *Data Protection Act 1998.* [online] Available at http://www.legislation.gov.uk/ukpga/1998/29/contents

Lewis, C.S. (1940). *The Problem of Pain.* London: HarperCollins.

Lewis, C.S. (1956). *The Last Battle*. London: HarperCollins.

LifeNaut Project (2017). Available at https://www.lifenaut.com/

Lifton, R. J. (1973). The sense of immortality: On death and the continuity of life. *American Journal of Psychoanalysis*, *33*, 3–15.

Linds, W. (2006). Metaxis: Dancing in the in-between. In J. Cohen-Cruz & M. Shutzman (Eds.), *A Boal Companion: Dialogues on Theatre and Cultural Politics*. New York: Routledge, 114–124.

Longhurst, R. (2001). *Bodies: Exploring Fluid Boundaries*. London: Routledge.

Lorenz, H. (2009). Ancient theories of soul. In E. N. Zalta (Ed.), *The Stanford Encyclopedia of Philosophy* (Summer 2009 Edition), Available at https://plato.stanford.edu/archives/sum2009/entries/ancient-soul/.

Lynn, C., & Rath, A. (2012). GriefNet: Creating and maintaining an internet Bereavement community. In C. Sofka, I. N. Cupit, & K. R. Gilbert (Eds.), *Dying, Death and Grief in an Online Universe*. New York: Springer.

Mbiti, J.S. (1990). *African Religions and Philosophy* (2nd edition). London: Heinemann.

McLaughlin, C., & Vitak, J. (2012). Norm evolution and violation on Facebook. *New Media & Society*, *14*(2), 299–315. doi:10.1177/1461444811412712

McDannell, C., & Lang, B. (2001). *Heaven – A History*. New Haven, CT: Yale University Press.

Mckinley, J.I. (1994). A pyre and grave goods in British cremation burials; have we missed something? *Antiquity*, *68*(258), 132–134.

Maciel, C., & Pereira, V. (2013). *Digital Legacy and Interaction*. Heidelberg: Springer.

Maddrell, A., & Sidaway, J. (2010). Introduction: Bringing a spatial lens to death, dying, mourning and remembrance. In A. Maddrell & J. Sidaway (Eds.), *Deathscapes: Spaces for Death, Dying and Bereavement*. Aldershot: Ashgate, 1–18.

Maddrell, A. (2016). Mapping grief: A conceptual framework for understanding the spatial dimension of bereavement, mourning and remembrance. *Social and Cultural Geography*, *17*, 166–188.

Mandela, N. (1994). *The Long Walk to Freedom*. London: Abacus.

Mao, F. (2018). 'Real bodies' exhibition causes controversy in Australia. 26th April. Available at https://www.bbc.co.uk/news/world-australia-43902524

Mason-Robbie (2021). Experience of using a virtual life coach: A case study of novice users. In M. Savin-Baden (Ed.), *Postdigital Humans: Transitions, Transformations and Transcendence*. Singapore: Springer.

Mayer-Schoenberger, V. (2009). *Delete: The Virtue of Forgetting in the Digital Age*. Princeton, NJ: Princeton University Press.

Meese, J., Nansen, B., Kohn, T., Arnold, M., et al. (2015). Posthumous personhood and the affordances of digital media. *Mortality*, *20*(4), 408–420.

Meier, E. A., Gallegos, J. V., Thomas, L. P., Depp, C. A., Irwin, S. A., & Jeste, D. V. (2016). Defining a good death (successful dying): literature review and a call for research and public dialogue. *The American Journal of Geriatric Psychiatry: Official Journal of the American Association for Geriatric Psychiatry*, *24*(4), 261–271. doi:10.1016/j.jagp.2016.01.135

Mellor, P., & Shilling, C. (1993). Modernity, self-identity and the sequestration of death. *Sociology*, *27*, 411–431.

Mey, J. L. (2006). Pragmatic acts. In K. Brown (Ed.), *Encyclopedia of Language and Linguistics (Online Version)*. Oxford: Elsevier.

Meyer, J. H. F., & Land, R. (2006). Threshold concepts and troublesome knowledge: Issues of liminality. In J. H. F. Meyer & R. Land (Eds.), *Overcoming Barriers to Student Understanding: Threshold Concepts and Troublesome Knowledge* (pp. 19–32). London and New York: Routledge.

Meyer, M., & Woodthorpe, K. (2008).The material presence of absence: A dialogue between museums and cemeteries. *Sociological Research Online*, *13*(5), 127–135.

Miles, W. (2002). Auschwitz: Museum interpretation and darker tourism. *Annals of Tourism Research*, *29*, 1175–1178.

Moltmann, J. (1996). *The Coming of God: Christian Eschatology*. London: SCM.

Morgan, D. (2013). Religion and media: A critical review of recent developments. *Critical Research on Religion*, *1*(3), 347–356.

Mori, J., Gibbs, M., Arnold, M., & Nansen, B. (2012). Design considerations for after death: Comparing the affordances of three online platforms. Proceedings of the 24th Australian Computer-Human Interaction Conference, 395–404. doi:10.1145/2414536.2414599

Morse, T. (2021). Now trending: #Massacre. On the ethical challenges of spreading spectacular terrorism on new media. In M. H. Jacobsen (Ed.), *The Age of Spectacular Death*. Oxford: Routledge.

Morse, T., & Birnhack, M. (2020). Digital remains: The users' perspectives. In M. Savin-Baden & V. Mason-Robbie (Eds.), *Digital Afterlife. Death Matters in a Digital Age*. Boca Raton, FL: CRC Press.

Mosco, V. (2004). *The Digital Sublime: Myth, Power and Cyberspace*. Cambridge, MA: The MIT Press.

Mowlabocus, S. (2016). The 'mastery' of the swipe: Smartphones, transitional objects and interstitial time. *First Monday*, *21*(10).

Moyer, L. M., & Enck, S. (2020). Is my grief too public for you? The digitalization of grief on Facebook™, *Death Studies*, *4*(2), 89–97.

Murray, M. (2010). Laying Lazarus to rest: The place and the space of the dead in explanations of near death experiences. In A. Maddrell & J. D. Didaway (Eds.), *Deathscapes. Spaces for Death, Dying, Mourning and Remembrance*. Oxfordshire: Routledge, 37–53.

Murray, S. J. (2006). Thanatopolitics: On the Use of Death for Mobilizing Political Life. *Polygraph: An International Journal of Politics and Culture*, *18*, 191–215.

Murray-Parkes, C. (1971). Psycho-social transitions: A field for study. *Social Science and Medicine*, *5*(2), 101–115.

Mwakilema, A. A. B. (1997). *Death and Life After Death in the Nyakyusa Belief.* Diploma in Theology thesis, St. Mark's Theological College, Dar es Salaam, Tanzania.

Nansen, B., Arnold, M., Gibbs, M., & Kohn, T. (2015). The restless dead in the digital cemetery, digital death: Mortality and beyond in the online age. In C. M. Moreman & A. D. Lewis (Eds.), *Digital Death: Mortality and Beyond in the Online Age*. Santa Barbara, CA: Praeger, 111–124.

Nansen, B., Kohn, T., Arnold, M., van Ryn, L., & Gibbs, M. (2017). Social media in the funeral industry: On the digitization of grief. *Journal of Broadcasting & Electronic Media*, *61*(1), 73–89. doi: 10.1080/08838151.2016.1273925

Nikishina, V. B., Sokolskaya, M. V., Musatova, O. A., Loskutova, I. M., Zapesotskaya, I. V., & Bogomolova, O. Y. (2020). The phenomenon of "digital" death: formation and genesis of the attitude to death in social networks of students. *Cypriot Journal of Educational Science*, 15(5), 1262 –1275. doi:10.18844/cjes.v15i5.5166

O'Connor, M. (2020). Posthumous digital material: does it 'live on' in survivors' accounts of their dead? In M. Savin-Baden & V. Mason-Robbie (Eds.), *Digital Afterlife. Death Matters in a Digital Age*. FL: Boca Raton, CRC Press.

O'Connor, M. & Kasket, E. (2021, in preparation). What grief isn't: Dead grief concepts and their digital-age revival. In T. Machin, C. Brownlow, J. Gilmour, & S. Abel (Eds.), *Social Media and Technology Across the Lifespan* (Chapter 8). London: Palgrave Macmillan.

Öhman, C. J., & Watson, D. (2019). Are the dead taking over Facebook? A big data approach to the future of death online. *Big Data & Society*. doi:10.1177/2053951719842540

Oliver, D.P., Washington, K.T., Wittenberg-Lyles, E., Demiris, G., & Porock D. (2009). 'They're part of the team': Participant evaluation of the ACTIVE intervention. *Palliative Medicine*, 23(6), 549–555.

Palmer, M. (2015). From grief police to tragedy hipsters: The seven most unhelpful arguments that took over the media after the Paris attacks. *Junkee*. Available at http://junkee.com/from-grief-police-to-tragedy-hipsters-the-seven-most-unhelpful-arguments-that-took-over-the-media-after-the-paris-attacks/69619

Papailias, P. (2016). Witnessing in the age of the database: Viral memorials, affective publics, and the assemblage of mourning. *Memory Studies*, 9(4), 437–454.

Parvaresh, V. (2017). Introduction: death, dying and the pragmeme. In V. Parvaresh & A. Capone (Eds.), *The Pragmeme of Accommodation: The Case of Interaction around the Event of Death*, Perspectives in Pragmatics, Philosophy & Psychology 13, Cham: Springer.

Perfect Choice Funerals. (2020). 70% of over 50s are unaware that social media accounts can be memorialised, leaving thousands of 'ghost' profiles active. Available at https://www.funeralplans.co.uk/in-the-news/70-of-over-50s-unaware-that-social-media-accounts-can-be-memorialised/

Pelletier, C. (2005). New Technologies, New Identities: The University in the Informational Age. In R. Land & S. Bayne (Eds.), *Education in Cyberspace*. London: RoutledgeFalmer.

Penfold-Mounce, R., & Smith, R. (2021). Resisting the grave: Value and the productive celebrity dead. In M. H. Jacobsen (Ed.), *The Age of Spectacular Death*. Oxford: Routledge.

Peters, M. A., Jandrić, P., & Hayes, S. (2021a). Biodigital philosophy, technological convergence, and new knowledge ecologies. *Postdigital Science and Education*. doi:10.1007/s42438-020-00211-7.

Peters, M. A., Jandrić, P., & Hayes, S. (2021b). Biodigital technologies and the bioeconomy: The global new green deal? *Educational Philosophy and Theory*. doi:10.1080/00131857.2020.1861938

Phillips, P., Schiefelbein-Guerrero, K., & Kurlberg, J. (2019). Defining digital theology: Digital humanities, digital religion and the particular work of the CODEC research centre and network. *Open Theology, 5,* 29–43.

Phillips, W. (2011). LOLing at tragedy: Facebook trolls, memorial pages and resistance to grief online. *First Monday, 16*(12). doi:10.5210/fm.v16i12.3168

Phillips, W. (2015). *This is why we can't have nice things: Mapping the relationships between online trolling and mainstream culture.* Cambridge, MA: MIT Press Books.

Pitsillides, S. Katsikides, S., & Conreen, M. (2009). *Digital Death, IFIP WG9.5 Virtuality and Society International Workshop on Images of Virtuality: Conceptualizations and Applications in Everyday Life,* April 23–24, Athens, Greece.

Pitsillides, S., Waller, M., & Fairfax, D. (2012). Digital death, the way digital data affects how we are (re)membered. In S. Warburton & S. Hatzipanagos (Eds.), *Digital Identity and Social Media.* Hershey: IGI Global, 75–90.

Plato. (360 BCE). Symposium. (B. Jowett trans.). Available at http://classics.mit. edu/Plato/symposium.html

Postman, P. (1993). *Technopoly: The Surrender of Culture to Technology.* New York: Vintage Books.

Pullman, P. (1997). *The Subtle Knife.* New York: Scholastic Point.

Rahner, K. (1965). *On the Theology of Death.* London: Burnes and Oates.

Rahner, K. (1975). Ideas for a theology of death. In *Theological Investigations* (Vol. 17). New York: The Crossroad Publishing Company.

Reader, J. (2020). Philosophical investigations into digital afterlife. In M. Savin-Baden & V. Mason-Robbie (Eds.), *Digital Afterlife. Death Matters in a Digital Age.* FL: Boca Raton, CRC Press, 127–142.

Reader, J., & Savin-Baden, M. (2021). Postdigital theologies: Technology, belief and practice. *Postdigital Science and Education.* doi:10.1007/s42438-020-00212-6

Reader, J. and Savin-Baden, M. (2022, Forthcoming). Divine assemblages (Perichoresis):The digital as Scapegoat (Pharmakon). In J. Reader & M. Savin-Baden (Eds.), *Postdigital Theologies.* Cham: Springer.

Retsikas, K. (2007). Being and place: Movement, ancestors, and personhood in East Java, Indonesia. *Journal of the Royal Anthropological Institute, 13,* 969–86.

Riek, L. D., & Howard, D. (2014). A code of ethics for the human-robot interaction profession (April 4). Proceedings of we Robot, 2014. Available at https://ssrn.com/abstract=2757805.

Roberston, B. R. (2020). On the threshold of tomorrow. *Oneing, 8*(1), 57–60.

Rohr, R. (2002). Grieving as sacred space. *Sojourners Magazine, 31,* 20–24. Available at www.sojo.net/index.cfm?action

Rossetto, K. R., Lannutti, P. J., & Strauman, E. C. (2014). Death on Facebook: Examining the roles of social media communication for the bereaved. *Journal of Social and Personal Relationships, 32*(7), 974–994.

Rothblatt, M. (2014). *Virtually Human: The Promise—and the Peril—of Digital Immortality.* New York: St Martins' Press.

Rowling, J.K. (2007). *Harry Potter and the Deathly Hallows*. London: Bloomsbury.

Rycroft, G. (2020). Legal issues in digital afterlife digital immortality platform. In M. Savin-Baden & V. Mason-Robbie (Eds.), *Digital Afterlife. Death Matters in a Digital Age*. FL: CRC Press, 127–142.

Sandman, L. (2005). *A Good Death: On the Value of Death and Dying*. Berkshire: Open University Press.

Savin-Baden, M. (2008). *Learning Spaces. Creating Opportunities for Knowledge Creation in Academic Life*. Maidenhead: McGraw Hill.

Savin-Baden, M. (2021a). Postdigital afterlife and the spiritual realm. Paper presented at Death Online Research Symposium IT University of Denmark 21-23 April.

Savin-Baden, M. (Ed.). (2021b). What are postdigital humans? In M. Savin-Baden (Ed.), *Postdigital Humans: Transitions, Transformations and Transcendence*. Singapore: Springer.

Savin-Baden, M., & Major, C. (2013). *Qualitative Research: The Essential Guide to Theory and Practice*. London: Routledge.

Savin-Baden, M. Burden, D., & Taylor, H. (2017). The ethics and impact of digital immortality. *Knowledge Cultures*, 5(2), 11–19.

Savin-Baden, M., & Burden, D. (2019). Digital immortality and Virtual Humans. *Journal of Post Digital Science and Education 1*, 87–103 (Published online July 2018).

Savin-Baden, M., & Mason-Robbie, V. (Eds.). (2020). *Digital Afterlife. Death Matters in a Digital Age*. Boca Raton, FL: CRC Press.

Schaeffer, F., Zeoli, B. (Producers), & Gonser, J. (Director) (1977). *How Should We Then Live* [Motion Picture]. Muskegon, MI: Gospel Films.

Schweitzer, A. (1907/1974). Overcoming death. In *Reverence for Life*. London: SPCK, 67–81.

Seale, C. (1998). *Constructing Death: The Sociology of Death and Bereavement*. Cambridge: Cambridge University Press.

Sebold, A. (2003). *The Lovely Bones*. London: Picador.

Senft, T. M., & Baym, N. K. (2015). What does the selfie say? Investigating a global phenomenon. *International Journal of Communication*, 9, 1588–1606.

Shakespeare, W. (1599/2008). *Julius Caesar* (1st edition). Oxford: The Oxford Shakespeare OUP.

Shakespeare, W. (1609/2008). *Hamlet. Prince of Denmark* (Reissue edition). Oxford: OUP.

Shakespeare, W. (1633/2008). *The Tragedy of King Richard III* (1st edition). Oxford: The Oxford Shakespeare OUP.

Shakespeare, W. (1611/2008). *The Tempest*. Oxford: The Oxford Shakespeare OUP.

Sherlock, A. (2013). Larger than life: Digital resurrection and the re-enchantment of society. *The Information Society*, 29(3), 164–176.

Sibbett, C. H., & Thompson, W. T. (2008). Nettlesome knowledge, liminality and the taboo in cancer and art therapy experiences: implications for learning and teaching. In R. Land, J. H. F. Meyer, J. Smith (Eds.), *Threshold Concepts within the Disciplines*. Rotterdam: Sense Publishers, 227–242.

Smith, M. (2016). Majority of people want to be cremated when they die. Available at https://yougov.co.uk/topics/lifestyle/articles-reports/2016/08/16/majo rity-people-want-be-cremated-when-they-die

Sofka, C. (2020a). The transition from life to the digital afterlife: Thanatechnology and its impact on grief. In M. Savin-Baden & V. Mason-Robbie (Eds.), *Digital Afterlife. Death Matters in a Digital Age*. Boca Raton, FL: CRC Press.

Sofka, C. (2020b). Netiquette regarding digital legacies and dealing with death, tragedy, and grief. Available at https://www.siena.edu/faculty-and-staff/ person/carla-sofka/

Sontag, S. (1990). *On Photography*. New York: Anchor.

Staley, E. (2014). Messaging the dead: Social network sites and theologies of afterlife. In C. M. Moreman & A. D. Lewis (Eds.), *Digital Death Mortality and Beyond in the Online Age*. Santa Barbara, CA: Praeger.

Stroebe, M., & Schut, H. (1999). The dual process model of coping with bereavement: rationale and description. *Death Studies*, 23, 197–224.

Stokes, P. (2015). Deletion as second death: The moral status of digital remains. *Ethics and Information Technology*, 17(4), 237–248.

Stone, P. (2006). A dark tourism spectrum: Towards a typology of death and macabre related tourist sites, attractions and exhibitions. *Tourism: An international Interdisciplinary Journal*, 54, 145–160.

Surinano M. (2016). Sheol, the tomb, and the problem of postmortem existence. *The Journal of Hebrew Scriptures*, 6(11), 1–32.

Than, K. (2013). Neanderthal burials confirmed as Ancient Ritual. *National Geographic*, 16th December. Available at https://www.nationalgeographic. com/culture/article/131216-la-chapelle-neanderthal-burials-graves

Thomas, R. S. (1993). Kneeling. In *Collected Poem 1945-1990*. London: Phoenix, 199.

The Star (2019). Selfiecide: These are the world's capitals for death by selfie. Available at https://www.thestar.com.my/tech/tech-news/2019/02/14/selfiec ide-these-are-the-worlds-capitals-for-death-by-selfie/

Tonkin, L. (1996). Growing around grief—another way of looking at grief and recovery. *Bereavement Care*, 15(1), 10.

Trafford, V. (2008). Conceptual frameworks as a threshold concept in doctorateness. In R. Land, J. Meyer, & J. Smith (Eds.), *Threshold Concepts within the Disciplines*. Rotterdam: Sense Publications, 273–288.

Trubshaw, B. (2003). *The Metaphors and Rituals of Place and Time – An Introduction to Liminality or Why Christopher Robin Wouldn't Walk on the Cracks*. (First published in *Mercian Mysteries*, No. 22, February 1995). Available at www.indigogroup.co.uk/edge/ liminal.htm.

Tuan, Y.-F. (1977). *Space and Place: The Perspective of Experience*. Minneapolis, MN: Minnesota Press.

Turner, V. (1995). *The Ritual Process: Structure and Antistructure*. New York: Aldine de Gruyter.

Turner, V. (1982). *From Ritual to Theatre: The Human Seriousness of Play*. New York: Performing Arts Journal Publications.

Tynan, D. (2016). Augmented Eternity; Scientists aim to let us speak form beyond the grave. *The Guardian*. June 23. Available at https://www.theguardian. com/technology/2016/jun/23/artificial-intelligence-digital-immortality-mit-ryerson

Van Gennep, A. (1909). *The Rites of Passage* (English trans 1960 M. B. Vizedom & G. L. Caffee). London: Routledge & Kegan Paul.

VanderLeest, S. H. & Schuurman, D. C. (2015). A Christian perspective on Artificial Intelligence: How should Christians think about thinking machines? Proceedings of the 2015 Christian Engineering Conference (CEC), Seattle Pacific University, June, 2015, 91–107.

Virilio, P. (1997). *Open Sky*. London: Verso.

Wachowski, L., & Wachowski, A. (1999). *The Matrix* [Film]. Burbank, CA: Warner Bros.

Wagner, R. (2012). *Godwired: Religion, Ritual and Virtual Reality*. New York: Routledge.

Wagner, A. J. M. (2018). Do not click "Like" when somebody has died: The role of norms for mourning practices in social media. *Social Media + Society*. doi:10.1177/2056305117744392.

Walker, P., & Lane, V. (2011). Ghost bikes: memorials to road victims blamed for putting people off cycling. *The Guardian*, 10th November. Available at https:// www.theguardian.com/lifeandstyle/2011/nov/10/ghost-bikes-memorials-cycling-victims

Walter, T. (1996). A new model of grief: Bereavement and biography. *Mortality*, *1*(1), 7–25.

Walter, T. (2000). Grief narratives: The role of medicine in the policing of grief. *Anthropology & Medicine*, *7*(1), 97–114. doi:10.1080/136484700109377

Walter, T. (2016). The dead who become angels: Bereavement and vernacular religion. *OMEGA: Journal of Death and Dying*, *73*(1), 3–28. doi:10.1177/0030222815575697

Walter, T. (2017). How the dead survive: Ancestors, immortality, memory. In M. H. Jacobsen (Ed.), *Postmortal Society. Towards as Sociology of Immortality*. London: Routledge.

Walter, T. (2019). The pervasive dead. *Mortality*, *24*(4), 389–404.

Walter, T., Hourizi, R., Moncur, W., & Pitsillides, S. (2012). Does the internet change how we die and mourn? An overview. *OMEGA: Journal of Death and Dying*, *64*(4), 275–302.

Whelan, J. (2008). Metaxis. *After the Future*. Available at http://afterthefuture.typ epad.com/afterthefuture/2008/12/metaxis.html

Windisch, P., Hertler, C., Blum, D., Zwahlen, D., Förster, R. (2020). Leveraging advances in Artificial Intelligence to improve the quality and timing of palliative care. *Cancers*, *12*(5), 1149. doi:10.3390/cancers12051149

Winnicott, D. W. (1953). Transitional objects and transitional phenomena. *The International Journal of Psycho-Analysis*, *34*, 89–97.

Winnicott, D. W. (1960). The theory of the parent-infant relationship. *The International Journal of Psycho-Analysis*, *41*, 585–595.

Wittgenstein, L. (1961). *Tractatus Logico-philosophicus*. London: Routledge & Kegan.

Woodthorpe, K. (2010). Buried bodies in an East London Cemetery: Revisiting taboo. In A. Maddrell & J. D. Didaway (Eds.), *Deathscapes. Spaces for Death, Dying, Mourning and Remembrance*. Oxfordshire: Routledge, 57–74.

Wright, D. W.M. (2021). Immersive dark tourism experiences. In M. H. Jacobsen (Ed.), *The Age of Spectacular Death*. Oxford: Routledge.

Wright, N. (2014). Death and the Internet: The implications of the digital afterlife. *First Monday*, *19*, Number 6 - 2 June 2014. Available at https://firstmonday.org/ojs/index.php/fm/article/download/4998/4088. doi:10.5210/fm.v19i6.4998

Žižek, S. (1999). The matrix, or two sides of perversion. In *Philosophy Today: Supplement on Extending the Horizons of Continental Philosophy*, *43*. Available at https://nosubject.com/The_Matrix,_or,_Two_Sides_of_Perversion

Zuboff, S. (2019). *The Age of Surveillance Capitalism: The Fight for a Human Future at the New Frontier of Power*. New York: PublicAffairs.

Glossary

Augmented eternity: Bridging the gaps between life by eternalizing digital identity.

Autonomous agent: A software programme which operates autonomously according to a particular set of rules. Sometimes used synonymously with Chatbots but also often used to describe the non-human agents within a simulation or decision-making system.

Avatar: The bodily manifestation of one's self or a virtual human in the context of a 3D virtual world or even as a 2D image within a text-chat system.

Bandwagon mourners: (Rosetto et al, 2014) Individuals who were only distant friends or associates of the deceased whilst they were alive but after their death post disproportionate commentary and pictures about them.

CemTech: Technologies that enhance the experience of visiting or interacting with a cemetery.

Chatbots: Software programmes which attempt to mimic human conversation when communicating with another (usually human) user. The Turing Test is a standard test of the maturity of Chabot technology. Chatbots may also be used to control 3D avatars within a virtual world, 2D avatars on a website or exist as participants within text-only environments such as chat rooms. Chatbots are conversational agents which support a wide range of natural language and extended conversations rather than just question and answer or command and response.

Conversational agents: Computer programs which use a natural language rather than a command line (or other) interface to a system, database or program.

Death goods: Artefacts such as coffins, urns and plaques that are sold by funeral directors and other companies.

Death Tech: The growth of technology connected with death and bereavement.

Digital afterlife: The idea of a virtual space, where information, assets, legacies and remains reside as part of the cyber soul.

Digital assets: Any electronic asset of personal or economic value.

Digital capabilities: The range of skills and understandings needed to operate digital media and collaborate and share with others, as well as being digitally literate and fluent.

Digital consciousness: The development of emotional and intellectual immortality through mind clones.

Digital death: Either the death of a living being and the way it affects the digital world or the death of a digital object and the way it affects a living being.

Digital estate planning: Organizing digital assets to ensure that these are handed over correctly to the deceased's dependents or loved ones.

Digital endurance: The creation of a lasting digital legacy and being posthumously present through digital reanimation.

Digital human: A software version of a human which is typically focused on the size and shape of the human for ergonomic research purposes.

Digital humanities: Applying computer-based technology to the humanities.

Digital inheritors: Those who inherit digital memories and messages following the death of a significant other.

Digital legacy: Digital assets left behind after death.

Digital mourning labour: This is an activity undertaken by corporate brands who use social media to share (and gain from) emotions of grief and nostalgia about dead celebrities.

Digital necromancy: The preservation and reanimation of digital remains.

Digital religion: The influence religion and new media have upon one another.

Digital remains: Digital content and data which were accumulated and stored online during the deceased's lifetime that reflect their digital personality and memories.

Digital resurrection: The use of dead people in media after death.

Digital theology: The use of the digital to study theology.

Digital traces: Digital footprints of the deceased left behind through digital media.

Emotional rubbernecking (DeGroot, 2014): Individuals who did not know the deceased while they were alive, but feel a connection with them (or their survivors) and visit and post on their memorial pages.

Grave goods: Items buried along with the body which include personal possessions and sometimes also artefacts, food and drink designed to ease the deceased's journey into the afterlife and provide for their needs.

Grief policing: A term of abuse for those who might presume to tell others how, or for whom, to grieve (Walter, 2000).

Media theology: The ideological dimension of the connection between religion and media.

Memorial trolling or RIP Trolling: (Phillips, 2015) One of the most distasteful forms of trolling. Perhaps one of the most famous RIP trolls was Sean Duffy, who was imprisoned for posting messages like 'Help me mummy, It's hot in Hell' on a dead girls' page on Mother's Day. RIP Trolling is often done anonymously or through the use of a pseudonym or fake identity.

Networked theology: The ways in which religion is practised online and offline in a networked society.

Online religion: The ways in which the Internet prompts new religious practices online.

Parasocial relationship: A one-sided relationship that media users form as a result of exposure to celebrities

Pedagogical agents: Virtual humans used for education purposes.

Photorealism: The original term refers to the genre of painting based on a photograph. It is used in digital technology to refer to the highly realistic reproduction of objects, peoples and environments such that they are indistinguishable from images of their real-life equivalents.

Posthumous personhood: The idea of a model of a person that transcends the boundaries of the body.

Post-mortem privacy: The right of the deceased to control his personality rights and digital remains post-mortem, broadly, or the right to privacy and data protection post-mortem.

Postmortal privacy: This is the protection of the informational body – the 'body' that keeps living post-mortem through the deceased's personal data, social networks, memes and other digital assets.

Post-mortem ownership: The ownership of assets – physical and digital, by inheritors post-death.

Presencing: In a social media context this refers to an individual situating themselves in place or space (such as at a funeral) and making the occasion known to others through social media by a comment, post, photograph or tweet.

Second death: The deletion of digital remains.

Second loss: The loss experienced due to the deletion of digital remains.

Spiritual realm: A realm that is connected to the physical world. It exists alongside it. Many people see the spiritual realm as something akin to heaven, or somewhere we go to post-death, and others see it as a place of ghosts, spirits and deities.

Symbolic immortality: (Lifton, 1979) The idea that individuals seek a sense of life continuity, or immortality, through symbolic means.

Technologically mediated mourning: The use of social networking sites to mourn and memorialize those who have died physically.

Text chat: The means of communicating by text message and specifically in immersive virtual worlds by typing a response to another avatar in the world rather than using voice. Text chat may be private, public or in a closed group.

Text-to-speech: The techniques and technology which can convert text (typically generated by a computer but possibly also scanned from a page of text) to audible speech.

Thanatechnology: Any kind of technology that can be used to deal with death, dying, grief, loss and illness.

Thanatopathia: The passion for violent entertainment that is focused on death so that death becomes a trendy spectacular commodity as well as popular entertainment.

Thanatopolitics: The exploration of the politics of death.

Virtual humans: Software programs which present as human and which have behaviour, emotion, thinking, autonomy and interaction modelled on physical humans.

Virtual humanoids: Simple virtual humans which present, to a limited degree, as humans and which may reflect some of the behaviour, emotion, thinking, autonomy and interaction of a physical human.

Virtual reality: A simulated computer environment in either a realistic or an imaginary world. Most virtual reality emphasizes immersion so that the user suspends belief and accepts it as a real environment and uses a head-mounted display to enhance this.

Virtual sapiens: Sophisticated virtual humans that are designed to achieve similar levels of presentation, behaviour, emotion, thinking, autonomy and interaction with a physical human.

Index

Printed in the United States
by Baker & Taylor Publisher Services